4

Illustrated
Encyclopedic Dictionary
of Electronic Circuits

Illustrated Encyclopedic Dictionary of Electronic Circuits

JOHN DOUGLAS-YOUNG

Prentice-Hall, Inc.
Business and Professional Division

Englewood Cliffs, New Jersey

Prentice-Hall International, Inc., *London*
Prentice-Hall of Australia, Pty. Ltd., *Sydney*
Prentice-Hall of Canada, Ltd., *Toronto*
Prentice-Hall of India Private Ltd., *New Delhi*
Prentice-Hall of Japan, Inc., *Tokyo*
Prentice-Hall of Southeast Asia Pte. Ltd., *Singapore*
Whitehall Books, Ltd., Wellington, *New Zealand*
Editora Prentice-Hall do Brasil Ltda., *Rio de Janeiro*

©1983, *by*

PRENTICE-HALL, INC.

Englewood Cliffs, NJ

Editor: George E. Parker

Library of Congress Cataloging in Publication Data
Douglas-Young, John.
   Illustrated encyclopedic dictionary of electronic
circuits.
   Includes index.
   1. Electronic circuits—Dictionaries.   2. Electronic
circuit design.   I. Title.
TK7867.D65   1983   621.381′53′0321   82-23067
ISBN 0-13-450734-7

PRINTED IN THE UNITED STATES OF AMERICA

How to Use This Book

This book is quite different from any other electronic circuit book you've ever seen. What makes it unique are the three, closely related parts into which it is divided:

Part I   Illustrated Encyclopedic Dictionary of Electronic Circuits

Part II   Designing, Breadboarding, and Final Assembly

Part III   Performance Testing and Troubleshooting

In addition, there is an **Appendix** containing schematic symbols, electronic formulas, mathematical tables, wire tables, and a wide range of other helpful data.

The following summary of the contents will immediately reveal this book's practical value for everyone interested in the design and construction of electronic projects, or in understanding how circuits operate. Also, you'll see that you won't require an engineering degree to understand it.

Part I: Illustrated Encyclopedic Dictionary of Electronic Circuits

The total number of circuits that have been invented since the electronic age began is beyond imagination. Most of these circuits are inaccessible in the archives of great corporations. Even if we could get at them we couldn't publish them all. Fortunately for us, we don't have to. Most circuits are variations, more or less elegant (often to get around some one else's patent), of a relatively small number of circuits. An amplifier is an amplifier is an amplifier! So, for this section I have selected a few hundred practical examples of circuits that have thousands of applications, alone or in combination. With only minor exceptions, each schematic diagram is accompanied by a description of its operation and a parts list with component values.

Some circuit books don't give parts values. Their argument is that there are too many variables, so that the circuit by a reader might not work. If the reader had gone strictly by the book he would blame it for his nonsuccess, and might even write a nasty letter to the author. This doesn't credit the reader with much intelligence, since if he or she has had any experience at all with electronics, he or she will know that electronics is an art, not a science, and that in it Murphy's Law is dominant, if not rampant.

Since Murphy's Law says that if a thing can go wrong it will, no competent engineer ever expects to score a bull's-eye with his first shot. In fact, this is why **breadboarding** was invented. Breadboarding is the "cut-and-try" procedure, where you build a prototype of the circuit to see if it works and make changes as necessary. Only when you are satisfied that all the bugs are out of it do you proceed with the final assembly. (Breadboarding is covered in Part II.)

So, you are given parts values as a starting point. Many circuits will work fine just as they are, because they are not critical, but some may require a little experimentation. Good luck!

The circuits given in this section are all organized in alphabetical sequence, making it extremely simple to find the one you want, whether for a project you're working on, or to help you in troubleshooting a defective piece of electronic gear. In addition, cross-references direct you to other versions, or to a major reference about that category of circuits. The major reference gives information in depth to help you design a new circuit, adapt an existing one, or combine two or more, with every expectation that the final version will be workable. This helps you to understand just what needs to be done to make defective equipment work properly.

The major references help to make this dictionary "encyclopedic." They are located alphabetically in the same A-Z listing as the smaller entries, and consist of in-depth guides that explain the basic principles underlying the circuits covered. This permits you to start from general information, and proceed to focus on a specific circuit, instead of the other way round.

The major references are articles about groups of circuits, such as Amplifiers, Integrated Circuits, Oscillators, Power Supplies, and so on. Each of the dictionary entries gives the name of the major reference pertaining to it in italics, and each major reference lists all the circuits in its category. In this way the two types of entry also act as cross references for each other.

For example, see the following dictionary entry.

Battery Eliminator (Figure B-9)

This is the circuit of a power supply used to substitute for a battery in order to power a transistor radio or other battery-operated device. Power transformer T1 converts the line voltage to 12.6 volts AC, which is then rectified by D1. The resulting positive half-cycles charge C1. Zener diode D2 breaks down at 10.2 volts, so this potential is applied to the base of Q1 regardless of what the voltage on C1 might be. When the line voltage decreases or a load is connected, so that the output voltage tends to decrease, the forward bias on Q1 increases and a greater current flows through the transistor. Conversely, any tendency for the output voltage to increase reduces the forward bias on Q1 and decreases the current it can pass.

You can get other outputs with a different T1 and D2. Substituting a 5-kilohm potentiometer for R1 and D2, with the sliding contact connected to Q1's base, will give you a variable power supply. See *Power Supply Circuits*.

Parts List

C1   1000 microfarads, 35 volts, electrolytic
D1   Silicon rectifier diode, 6000 volts, 1 ampere
D2   Zener diode, 10.2 volts
Q1   Silicon NPN power transistor, 2N3393, or equivalent
R1   220 ohms, 1 watt
T1   Power transformer, 117/12.6 volts AC, 1 ampere

This might be all the information you want. However, after reading this entry, you might conclude you'd prefer a different type of power supply. In this case you'd turn to the major reference, *Power Supplies,* where you'd find it among the 27 power-supply types listed. If you'd already decided on a specific type, such as a *laboratory power supply,* you could turn to it directly. The major reference, *Power Supplies,* explains how AC is converted to DC. You are told about power transformers, rectifiers, filters, regulation, voltage doublers, triplers, and much more. This information makes it easy for you even to design your own power supply, because *all* of the background data you need is there.

Each of the other major references tells you what you want to know to design your own circuit, and lists other circuits in the same category that are given elsewhere in the encyclopedia. You have the choice of custom-made or ready-made.

This section of the book, combining quick reference with in-depth knowledge, is a powerful design tool that serves as a unique guide to understanding hundreds of electronic circuits.

Part II: Designing, Breadboarding, and Final Assembly

When you have selected a circuit to build from Part I, you have already performed an important part of its design, but there is much more to do before its schematic diagram is translated into actual hardware. Part II tells you how by detailing in words and pictures the step-by-step procedure of the fabrication from scratch of a typical electronic device. Starting with a "shopping list" of all the component parts required, it tells you how to make the experimental layout, or breadboard. Then it tells you how to check the actual performance of the physical circuit. This is when you make any necessary changes, and at the same time determine the best arrangement of the parts for permanent installation. The way you do this is a necessary preliminary to the selection of a cabinet or enclosure of the proper size or type. At the end of this stage, you go on to the final assembly.

This whole section will simply such things as wire wrapping, making printed circuit boards, using "perfboards," drilling the front panel for controls and meters, dressing up the finished cabinet with decals, knobs, and so on, so that the completed project has a truly professional look.

Part III: Performance Testing and Troubleshooting

Performance testing refers to measuring parameters, or verifying operation, to see whether the equipment complies with the specifications it was designed to meet. You are taken through a step-by-step procedure, using the same typical device described in Part II, to demonstrate the way to do this with any type of equipment.

Troubleshooting means finding out what is wrong when some device fails to perform as it should. It does not actually include the repair procedure, but when the cause of the malfunction has been located it is obvious what has to be done.

There are dependable ways to troubleshoot, drawn from the experience of thousands of engineers and technicians over the years, that have proved to be more efficient than others. These tricks of the trade are explained here. You will find this section very valuable, because it shows you how to pinpoint the trouble in the least amount of time by using a logical approach that makes sense.

The *Illustrated Encyclopedic Dictionary of Electronic Circuits* offers an unusually helpful format with extraordinary value to technicians, engineers and experimenters alike. For practical utility, fast reference, and dependable data, it is a must for all who are active in the field of electronics.

John Douglas-Young

CONTENTS

PART I

Illustrated
Encyclopedic Dictionary

Absorption Modulator (Figure A-1)

This is a simple way of amplitude-modulating the radio frequency (RF) output of a low-power, continuous-wave radio transmitter. C1, L1, and L2 form the output tank circuit, connected by a suitable transmission line to the antenna E1. A link coil L3 of only one or two turns is placed near the tank coil L1. This link coil is connected to a carbon microphone MK1. Together they form a circuit that absorbs power from the transmitter's output. The amount of power

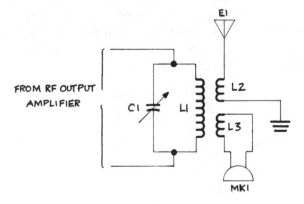

Figure A-1
Absorption Modulator

absorbed depends upon the resistance of MK1, which varies according to the sound waves impinging upon it. The signal radiated by the antenna, therefore, varies in amplitude in conformity with the sound waves.

This is an inefficient method of modulation, since it reduces the output power. It can only be used with a transmitter with an output of less than one watt, because MK1 is excited by the RF output power and could be damaged by anything higher. The output tank will also require retuning to counteract the detuning effect of the link coil circuit. See *Modulation Circuits*.

Parts List

MK1   Carbon microphone (as in telephone handset)

Absorption Frequency Meter (Figure A–2)

This device (also called a wavemeter) consists of a resonant circuit L1–C1 connected in series with a crystal detector D1 and a 0–1 milliammeter M1. L1 also serves as the pick-up coil, and D1 is tapped down on the inductance to improve the sensitivity and selectivity of the meter.

A set of four pick-up coils, constructed as described in the Parts List, will cover the range from 1.22 to 165 megahertz (MHz). These should be made to plug into a suitable socket mounted on the outside of the unit.

The meter operates by extracting a small amount of energy from the oscillating circuit to be measured. The frequency is then determined by tuning the frequency-meter circuit to resonance, and reading the frequency from a calibrated scale. Resonance is indicated by maximum deflection of the milliammeter pointer.

The unit is calibrated by means of a signal generator and a calibrated receiver. Set the receiver to a given frequency, tune the signal generator to the same frequency, and adjust the meter to resonance with the signal generator. This gives one calibration point. Repeat the procedure for other frequencies. When a sufficient number of calibration points has been obtained, you can draw a graph to show frequency *versus* dial settings on the frequency meter, or make a new dial with the frequency settings marked directly on it.

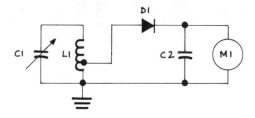

Figure A–2
Absorption Frequency Meter

Although the absorption frequency meter should not be depended upon for accurate measurement, it is a highly useful instrument to have in an amateur radio station, even when better frequency-measuring equipment is available. Since it generates no harmonics itself, it will respond only to the frequency to which it is tuned. It is therefore indispensable for distinguishing between fundamental and various harmonics, and for detecting harmonic and parasitic oscillations, RF in undesirable places, making rough measurements of field strength in adjustment of antennas, and so on. See *Test and Indicating Circuits*.

Parts List

C1   140 picofarad variable capacitor
C2   0.0015 microfarad mica capacitor
D1   Diode type 1N34 or equivalent
L1   *1.22–4.0 MHz:* 70 turns of No. 32 enameled wire, wound on a 1-inch diameter coil form, to make a coil 5/8 inch long, with a tap 12-1/2 turns from the grounded end.

　　　　4.0–13.5 MHz: 20 turns of No. 20 enameled wire, wound on a 1-inch diameter coil form, to make a coil 9/16 inch long, with a tap 4-1/2 turns from the grounded end.

　　　　13.2–44.0 MHz: 5 turns of No. 20 enameled wire, wound on a 1-inch diameter coil form, to make a coil 5/16 inch long, with a tap 1-1/2 turns from the grounded end.

　　　　39.8–165 MHz: Hairpin loop of No. 14 wire, 2 inches long. This length includes the ends which fit down into the coil-form prongs. The loop is fitted inside a coil form, and the wires are spaced 1/2 inch apart. The tap is 1-5/8 inches from the grounded end.

M1   Panel meter, 0–1 mA DC (Radio Shack Catalog Number 270–1752, or equivalent).

AC Adapter (Figure A–3)

This is the circuit used in most of the alternating current (AC) adapters supplied with pocket calculators and the like. A small transformer T1 reduces the line voltage (117 VAC at 60 Hz) to the desired output voltage, which is then rectified by D1. Since the pulsating DC that results is used to charge the batteries in the device, filtering is not provided. See *Power Supply Circuits*.

Parts List

D1   Rectifier diode, 1N4000 series, 1 ampere
T1   Miniature, 300 mA, 6.3, 12, or 24 VAC output as required

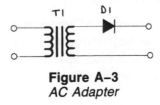

Figure A–3
AC Adapter

AC Ammeter (Figure A–4)

In this circuit a DC milliammeter M1 is used as an AC ammeter by rectifying the AC and measuring the resultant DC. T1 is a power transformer, with its secondary winding connected in series with the load. The latter is plugged into the AC output receptacles, while the input receptacle is plugged into the wall socket.

When the load current is zero no voltage is induced in the primary, so the meter pointer is not deflected. However, when a load is connected and load current flows through T1's secondary, a voltage is induced in the primary. This AC voltage is rectified by D1, and the resultant DC pulses charge C1 to a potential that is measured by the voltmeter consisting of R1 and M1.

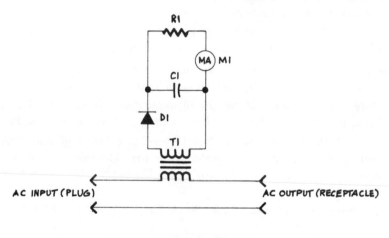

Figure A–4
AC Ammeter

Since T1 has a 120-VAC primary, and a 12-VAC secondary with a rating of 5 amperes, the AC potential across its primary will be 120 VAC when the load current is 5 amperes. When the load current is lower the primary voltage will be lower. The DC voltage across C1 will be 1.414 times the AC voltage across T1's primary; therefore, it will be 170 VDC when the load current is 5 amperes. Maximum deflection of the meter pointer is given by a current of 0.001 ampere, so the value of R1 must be such that the combined resistance of R1 and the meter divided into 170 VDC will give 0.001 ampere. The meter resistance is small enough to be ignored; therefore, R1 should have a value of 170 kilohms.

This device is not as accurate as a regular AC ammeter, but it is definitely less expensive, especially if you already have some of the parts on hand. Adjustment of the value of R1 can also improve its accuracy. A transformer with a higher secondary rating would allow measurement of higher load currents. See *Test and Indicating Circuits*.

Parts List

C1   1.0 microfarad, 250 WVDC
D1   Rectifier diode, 1N4000 series, 1A, 200V, or equivalent
M1   Panel meter, 0–1 mA, DC
R1   Resistor, 170 kilohms, 1/4 watt (or three 56-kilohm resistors in series)
T1   Power transformer, primary 120 VAC at 60 Hz, secondary 12 VAC, rated for 5
       amperes (other ratings can be used, of course)

AC-DC Power Supply

See *Universal AC-DC Power Supply.*

Active Detector (Figure A–5)

This circuit has an advantage over conventional diode detectors in that it provides a voltage gain of 3 instead of a loss. This is because Q1 and Q2 form a differential amplifier (q.v.). The output of Q2 is applied to the emitter follower Q3, which in turn provides both the audio output and an AGC voltage from the junction of R3 and R5. See *Demodulation Circuits.*

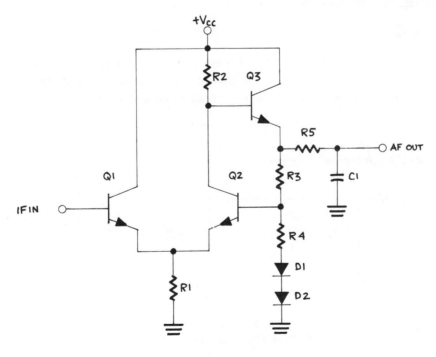

Figure A–5
Active Detector

Parts List

C1	0.01 microfarad, 50 WVDC
D1, D2	1N34A or equivalent
Q1, Q2, Q3	General-purpose RF/IF amplifier NPN transistors
R1	3 kilohms, 1/4 watt
R2	6.8 kilohms, 1/4 watt
R3	4 kilohms, 1/4 watt
R4	2 kilohms, 2 kilohms, 1/4 watt
R5	500 ohms, 1/2 watt

ACTIVE FILTER CIRCUITS

Filters are networks of electronic components that tailor signals to meet requirements of band-pass, phase shift, or time delay. There are passive filters (q.v.) and active filters. An active filter consists of an operational amplifier with an external network of resistors and capacitors. This network determines the function of the filter, whether it be band-pass, bandstop, constant time delay, high pass, low pass, or phase shift.

Low-pass filter

In this article we shall use the designing of a low-pass filter to show how any of the above-named filters is designed. Only "rule-of-thumb" methods will be used, but these will be as accurate for practical purposes as any advanced mathematical calculations.

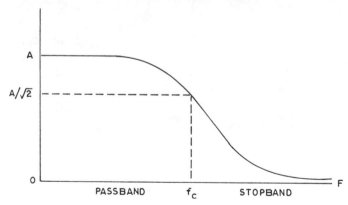

Figure A–6
Low-Pass Active Filter Response Curve

A low-pass filter passes lower frequencies but not higher ones, as in Figure A-6, where OF represents increasing frequency and OA increasing amplitude. The response of the filter is shown by the curve AF, which indicates that frequencies below f_c are passed readily, but that those above f_c are greatly attenuated. The range of frequencies lower than f_c is therefore called the passband, and the range higher than f_c is the stopband. The cutoff frequency is defined as the frequency at which the amplitude A has fallen to 0.707A, or 3 dB down.

Figure A-7 gives the circuit of a commonly-used low-pass active filter. It will have a gain of 2 or 10 according to the values of the components selected. For a gain of 2

    R1 = 27 kilohms
    R2 = 51 kilohms
    R3 = 33 kilohms

For a gain of 10 the values would be

    R1 = 15 kilohms
    R2 = 168 (100 + 68) kilohms
    R3 = 47 kilohms

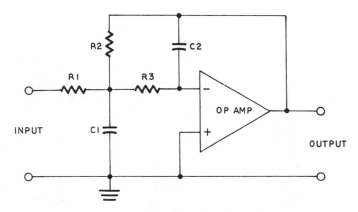

Figure A–7
Low-Pass Active Filter Circuit

The value of C1 in microfarads is obtained by dividing f_c into 10. For example, if f_c were 1000, C1 would be 10/1000 = 0.01 microfarad.

The value of C2 is found by multiplying 0.150 by the value of C1 for a gain of 2, or multiplying 0.033 by the value of C1 for a gain of 10. (The value of C1 is the same for either gain.)

The operational amplifier IC used is the popular bipolar 741 type or the FET 536 type, or equivalent.

Other Active Filter Circuits

See the following list for other active filter circuits described elsewhere in this encyclopedia:

    Band-pass active filter
    Bandstop active filter
    Constant time-delay active filter
    High-pass active filter

AGC

See *Automatic Gain Control*.

AM Detector and Audio Amplifier
(Figure A–8)

This circuit uses a dual-diode-triode tube as a combined AVC rectifier, second detector, and first AF amplifier. It was the standard circuit for the popular five-tube radios in use before transistors took over.

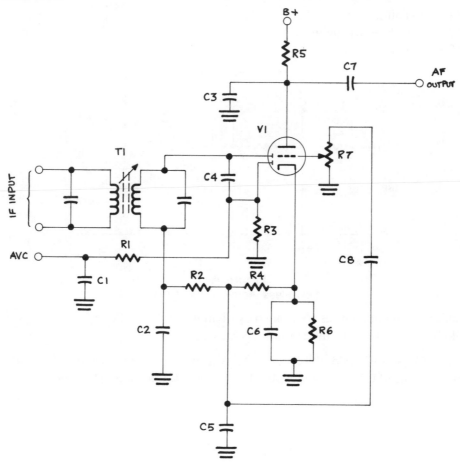

Figure A–8
AM Detector and Audio Amplifier

T1 is the final intermediate frequency (IF) transformer, through which the IF signal is coupled to one of the diode plates in V1. The potential on this plate is alternately positive and negative as the signal swings from one polarity to the other. Current can flow from the cathode to the plate only when the potential is positive, however, so that a series of positive half-cycles of current flow in the load circuit consisting of R2, R4, C2, and C5. In this circuit the resistances and capacitances are so proportioned that the capacitors charge to the peak value of the rectified voltage on each pulse, and retain enough charge between pulses so that the voltage across the resistors is smoothed out. C2 and C5 thus act as a filter for the RF component of the output of the detector, leaving a DC component that varies in the same way as the modulation on the original signal. When this varying DC voltage is applied to a following amplifier through the coupling capacitor C8, only the *variations* in voltage are transferred, so that an AC signal appears across the volume control R7. Adjustment of this control causes a voltage corresponding to the degree of adjustment to appear on the grid of the triode section of V1.

This is the audio amplifier stage of the circuit. R6 is the cathode bias resistor and C6 is the cathode bypass capacitor. Voltage output is developed across the load resistor R5 and coupled through C7 to the power amplifier (not shown).

The AVC circuit is discussed under automatic volume control (q.v.). See *Amplifier Circuits; Demodulation Circuits*.

Parts List

C1	0.1 microfarad
C2	100 picofarads
C3	270 picofarads
C4	100 picofarads
C5	100 picofarads
C6	5 to 10 microfarads, electrolytic
C7	0.01 to 0.1 microfarad
C8	0.01 to 0.1 microfarad
R1	0.5 to 1 megohm
R2	50 to 250 kilohms
R3	2 to 5 megohms
R4	270 kilohms
R5	250 kilohms
R6	1800 ohms
R7	500 kilohms variable
T1	IF transformer
V1	12AT6 or equivalent

Amplified S-Meter (Figure A–9)

This circuit enables you to see an inexpensive milliammeter as an S-meter in a radio receiver. A single transistor is used to boost the weak AVC signal to a level where it will give full-scale deflection on a 0–1 milliammeter. At the same time the relatively high input resistance of the amplifier has very little effect on the AVC operation of the receiver.

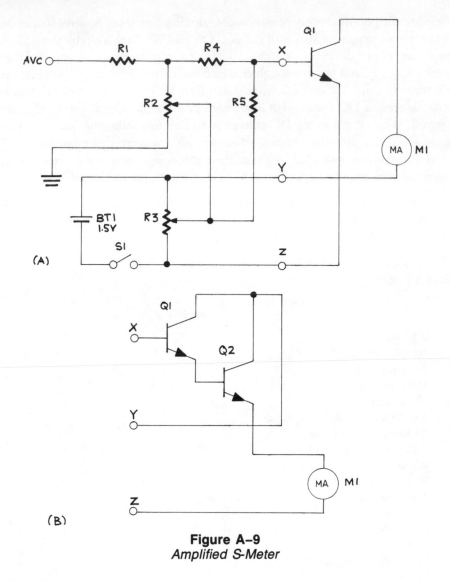

Figure A–9
Amplified S-Meter

The potentiometer R3 is used to set the meter to zero with no signal applied, while the other potentiometer R2 adjusts the sensitivity to give full-scale deflection with the strongest signal.

If more isolation and sensitivity are required, two transistors may be substituted, as shown in (B). They are connected to X, Y, and Z in place of the single transistor shown in (A). See *Test and Indicating Circuits*.

Parts List

BT1	1.5–V "D" cell
M1	0–1 mA DC panel meter

Q1, Q2 General-purpose, low-frequency amplifier NPN transistors
R1   12 kilohms
R2   750 kilohms, variable
R3   25 kilohms, variable
R4   10 kilohms
R5   10 kilohms
S1   SPST switch, any type

AMPLIFIER CIRCUITS

At the end of this article is a long list of amplifier circuits of many kinds that are described elsewhere in the encyclopedia, so that you can use one of them ready-made, as it were. If you want to design a transistor amplifier from scratch, you can use the following simple "rule-of-thumb" procedure.

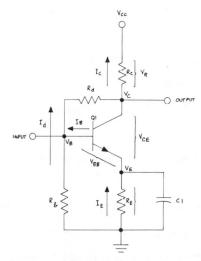

Figure A–10
Transistor Amplifier

This approach is usually just as good as the more elegant mathematical methods taught in school. The parameters of average transistors vary quiet a bit from one to another, so exact results cannot be expected. Instead, it is better to use lots of negative feedback to minimize transistor variability. The circuit in Figure A–10 does just this. It consists of a low- or medium-frequency voltage amplifier, using a general-purpose, small-signal transistor in a common-emitter configuration. The first thing we have to do is to get the following **parameters** for the transistor we are using:

Maximum allowable collector power dissipation (P_C)

Forward short-circuit current amplification factor (h_{fe}) or **current gain (beta)**, which are both assumed to be the same thing for our present purpose.

Then we shall also assume that all germanium transistors have an **emitter-base (DC) voltage (V_{EB})** of 0.2 volt, and that for all silicon transistors it is 0.7 volt.

Rule-of-thumb No. 1: Determine the collector-emitter voltage (V_{CE}) from:

$$V_{CE} = 0.6 \, V_{CC}$$

Rule-of-thumb No. 2: Determine the collector current (I_C) from:

$$I_C = P_C/V_{CE}$$

Rule-of-thumb No. 3: Determine the emitter current (I_E) from:

$$I_E = I_C \text{ (assume alpha = 1)}$$

Rule-of-thumb No. 4: Determine emitter voltage (V_E) from:

$$V_E = V_{EB}/0.1$$

Rule-of-thumb No. 5: Determine value of emitter resistance (R_E) from:

$$R_E = V_E/I_E$$

Rule-of-thumb No. 6: Determine value of external collector resistance (R_C) from:

$$R_C = 0.8 \left[\{ (V_{CC} - V_{CE})/I_E \} - R_E \right]$$

Rule-of-thumb No. 7: Determine base voltage (V_B) from:

$$V_B = V_E + V_{EB}$$

Rule-of-thumb No. 8: Determine base current (I_B) from:

$$I_B = I_C/h_{fe}, \text{ or } I_C/\text{beta}$$

Rule-of-thumb No. 9: Determine current (I_d) flowing in resistances R_d and R_b from:

$$I_d = I_B/0.1$$

Rule-of-thumb No. 10: Determine collector voltage (V_C) from:

$$V_C = V_{CE} + V_E$$

Rule-of-thumb No. 11: Determine combined resistance of R_d and R_b from:

$$(R_d + R_b) = V_C/I_d$$

Rule-of-thumb No. 12: Determine resistance of R_b from:

$$R_b = (V_B/V_C)(R_d + R_b)$$

Rule-of-thumb No. 13: Determine resistance of R_d from:

$$R_d = (R_d + R_b) - R_b$$

Rule-of-thumb No. 14: Determine voltage drop (V_R) across R_C from:

$$V_R = V_{CC} - V_C$$

Rule-of-thumb No. 15: Determine voltage gain (G_V) from:

$$G_V = 38.4\, V_R$$

Example: Let's suppose you want to design an amplifier using a germanium transistor with the following parameters:

$$P_C = 40\text{ mW}$$
$$H_{fe} = 50$$

Your first consideration is what collector power supply (V_{CC}) to use. As this is a voltage amplifier, you would like to develop as much voltage across R_C as possible without getting too high an I_C, for this would result in excessive P_C. Let's take 20 volts as a ballpark figure, and see how it works out. The calculations would go like this:

1. $V_{CE} = 0.6 \times 20 = 12$ volts
2. $I_C = 40/12 = 3.3$ milliamperes
3. $I_E = 3.3$ milliamperes
4. $V_E = 0.2/0.1 = 2$ volts
5. $R_E = 2/3.3 \times 10^{-3} = 600$ ohms (Use the nearest standard value, which is 560 ohms.)
6. $R_C = 0.8\, [\{(20–12)/3.3 \times 10^{-3}\} - 600]$
   $= 1459.4$ ohms (Use the nearest standard value, which is 1.5 kilohms.)
7. $V_B = 2 + 0.2 = 2.2$ volts
8. $I_B = 3.3 \times 10^{-3}/50 = 0.066 \times 10^{-3}$ ampere
9. $I_d = 0.066 \times 10^{-3}/0.1 = 0.66 \times 10^{-3}$ ampere
10. $V_C = 12 + 2 = 14$ volts
11. $(R_d + R_b) = 14/0.66 \times 10^{-3} = 21,212$ ohms
12. $R_b = (2.2/14)\,(21,212) = 3333$ ohms (Use the nearest standard value, which is 3.3 kilohms.)
13. $R_d = 21,212 - 3333 = 17,879$ ohms (Use the nearest standard values, which are 15 kilohms and 3.3 kilohms in series.)
14. $V_R = 20 - 14 = 6$ volts
15. $G_V = 38.4 \times 6 = 230$

The assumption of a V_{CC} of 20 volts turned out to be a good one, since you were able to get a satisfactory gain while staying within the allowable P_C.

The emitter capacitor should have a value of 100 microfarads to allow the emitter to be grounded for all AC frequencies.

Other Amplifier Circuits

See the following list for other amplifier circuits described elsewhere in the encyclopedia:

AM detector and audio amplifier
Audio amplifier with IC
Audio amplifier with JFET
Audio amplifier with triode
Audio output power splitter
Audio power output stage with IC driver
Audio power preamplifier
Audio preamplifier with operational amplifier
Automatic gain control
Automatic level control
Automatic volume control
Bass tone control
Broadband RF amplifier
Cascode JFET preamplifier
Cascode RF amplifier
Cathode-coupled amplifier
Cathode follower
Class-A amplifier (see Bias Circuits)
Class-B amplifier (see Bias Circuits)
Class-C amplifier (see Bias Circuits)
Common-base transistor amplifier
Common-collector transistor amplifier
Common-emitter transistor amplifier
Compression amplifier
Constant-gain audio amplifier
Current and voltage feedback amplifier
Darlington amplifier
Darlington differential amplifier
Degenerative neutralization
Differential amplifier
Direct-coupled amplifier
Filament-type tube audio amplifier
Frequency-compensated amplifier
Frequency-compensated preamplifier
Frequency-modulated IF amplifier with IC
Gain controls
Grounded-anode amplifier (see Grounded-Plate Amplifier)
Grounded-base amplifier (see Common-base Transistor Amplifier)
Grounded-cathode amplifier
Grounded-grid amplifier

Grounded-grid DC amplifier
Gounded-plate amplifier
Grounded-plate, grounded-grid amplifier
Hazeltine neutralization
IC audio-frequency amplifier
IC audio-frequency preamplifier
IC FM IF amplifier
IC IF amplifier (see AM Radio Using IC, Frequency-modulated IF Amplifier with IC, TV
    IF Amplifier with IC)
IC intercom
IC power amplifier
IC stereo preamplifier
IF amplifier (see IC IF Amplifier)
IF amplifier with ceramic filter
Inverter
Keyed AGC
Linear amplifier
Loudness control
Low-level audio amplifier
Magnetic amplifier
Neutralized push-pull triode RF amplifier
Neutralized transistor amplifier
Neutralized triode RF amplifier
Passive audio-frequency signal mixer
Phase inverter
Pi-network coupler
Preamplifier with operational amplifier
Push-pull amplifier (see Audio Power Output Stage, RF Amplifier with Two Triodes in
    Push-pull, Transformer-coupled Push-pull Power Amplifier)
RC-coupled amplifiers
Reflex amplifier
RF amplifier with IGFET
RF amplifier with one triode and neutralization
RF amplifier with transistor
RF amplifier with two triodes in push-pull
RF power amplifier with two triodes in push-pull (see RF Amplifier with Two Triodes in
    Push-pull)
Rice neutralization
Shunt degeneration
Shunt-fed triode power amplifier
Speaker combiner
Speaker splitter
Three-channel stereo synthesizer
Tone control

Transducer amplifier
Transformer-coupled mixer-balun
Transformer-coupled push-pull power amplifier
Transformer-coupled triode amplifier
Transistor amplifier
Transistor IF amplifier
Transistor output amplifier
Treble tone control
TV IF amplifier with IC
Video amplifier

Amplitude Modulation

See *Modulation Circuits*.

AM Radio Using Integrated Circuit (Figure A-11)

When would you want to use an integrated circuit as the basic subsystem of a typical low-cost AM radio? Economic considerations make it difficult for ICs to compete with discrete devices in these minimal-performance receivers. Only in high-performance console or communications receivers would the savings in the number of components required make it worthwhile.

Figure A–11 shows the circuit for an AM receiver using an IC. Double-tuned transformer-coupled circuits are used here, but any of the other forms of band-pass filters may be used. In addition to commerical broadcast receivers, amateur receivers would be a natural for the IC, as would hand-carried receivers.

The RF stage may take on any of several forms, and so is drawn only as a gain block. It could very well be another IC. Figure A–12 shows the internal subcircuits of the IC. The zener diode regulates the supply voltage to the biasing circuit. See *Integrated Circuits*.

Parts List

A1   RF stage (see text)
A2   RCA–CA3088E, or equivalent
C1   Antenna tuning capacitor (depends on band), ganged with C4 and C5
C2   0.1 microfarad
C3   1 microfarad
C4   RF stage tuning capacitor (see C1)
C5   Local oscillator tuning capacitor (see C1)
C6   0.01 microfarad
C7   0.05 microfarad
C8   0.05 microfarad
C9   (Internal in T3)

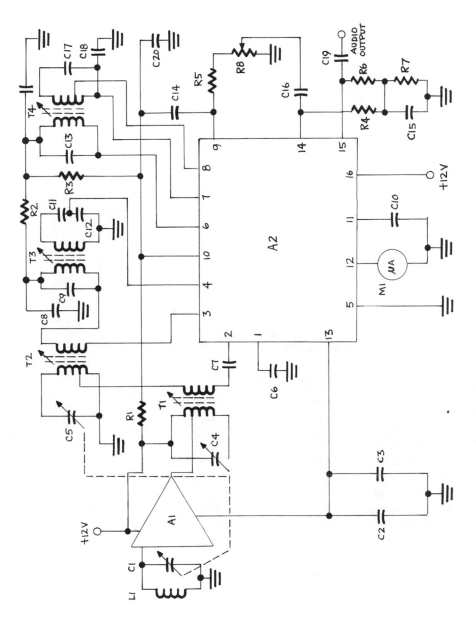

Figure A–11
AM Radio Using Integrated Circuit

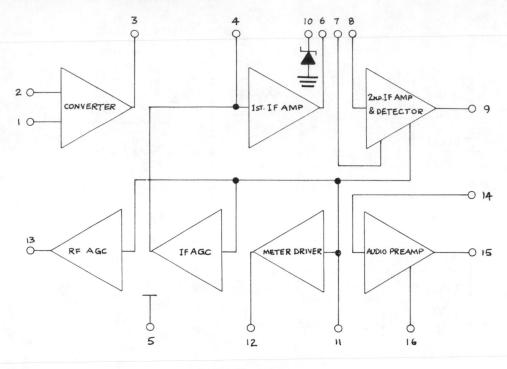

Figure A-12
Internal Subcircuits of A2 in Figure A-11

C10	50 microfarads, electrolytic
C11	130 picofarads
C12	0.039 microfarad
C13	(Internal in T4)
C14	0.02 microfarad
C15	10 microfarads, electrolytic
C16	25 microfarads, electrolytic
C17	(Internal in T4)
C18	0.05 microfarad
C19	25 microfarads, electrolytic
C20	0.1 microfarad
L1	Antenna tuning coil (depends on band)
M1	200-microampere DC meter (dial may be calibrated for 0–2 AGC volts, or S units)
R1	330 ohms
R2	200 ohms
R3	100 ohms
R4	5600 ohms
R5	2000 ohms
R6	6800 ohms
R7	1500 ohms
R8	10 kilohms, variable ("volume control")

T1	RF stage tuning transformer
T2	Oscillator tuning transformer
T3	Input IF transformer
T4	Output IF transformer

AND Gate (Figure A–13)

This circuit consists of two transistor switches in series. Current can flow between ground and V_{cc} only when both switches are on. Q1 and Q2 are turned on when a positive voltage is applied to their bases. In this condition, their very low resistance permits virtually all the supply voltage to appear across R3, so that the output goes positive as well. This basic gate is called an RTL (resistor-transistor-logic) AND gate. See *Logic Circuits*.

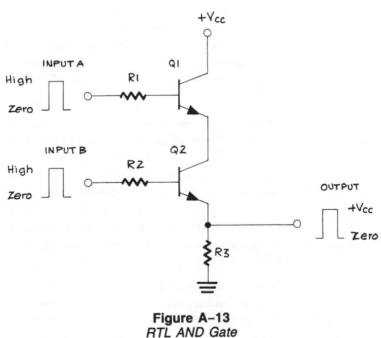

Figure A–13
RTL AND Gate

Parts List

Q1, Q2	Any general purpose or switching transistor
R1, R2	470 ohms
R3	640 ohms

ANTENNA CIRCUITS

These are circuits between the antenna and the transmitter or receiver which it serves. The essential consideration is that you are dealing with radio frequencies, so that impedance matching is of vital importance.

The transmission line between the antenna and the set has a characteristic impedance resulting from the capacitance and inductance that exist between the conductors. This has nothing to do with their resistance, which can be neglected in most cases.

Both the antenna and the set must have the same output or input impedance as the line. In this way a complete transfer of power takes place from the antenna to the line, and from the line to the receiver, or vice versa in the case of a transmitter. If the impedances do not match, some of the power is not absorbed. It is reflected back along the line and wasted.

There are two main types of transmission line. One is the parallel-conductor line, typified by the familiar twin lead used with TV antennas. The other is the coaxial cable, in which a solid or stranded-wire inner conductor is surrounded by polyethylene dielectric, over which an outer conductor of copper braid is woven, with the whole enclosed in a waterproof vinyl cover.

Most TV and FM antennas have an impedance of 300 ohms, and TV and FM receivers also have an input impedance of 300 ohms. Since twin lead has a characteristic impedance of 300 ohms, and is also inexpensive, it is widely used as a down lead. In urban strong-signal areas it presents few problems, but in other areas many people are using coaxial cable because the outer conductor provides shielding from noise pickup or signal loss due to the proximity of grounded metal objects.

The type of "coax" (RG–59/U) most commonly used, however, has a characteristic impedance of 75 ohms. To avoid mismatch it is necessary to make the antenna and receiver "think" they are connected by a 300-ohm line. This is done by installing a small matching transformer at each end of the line, so that the antenna and the receiver each has a 300-ohm load, while the line has a 75-ohm load at both ends. There is one other type of cable available in electronic retail outlets that combines the advantages of both twin lead and coax. This is 300-ohm shielded cable, consisting of twin lead surrounded by a polyfoam jacket, over which is laid a foil shield, which must be grounded. It requires no matching transformers but is somewhat stiffer than coax, so it is harder to handle.

RG–59/U coax is equally capable of connecting a transmitter to its antenna. However, in this case the transmitter will usually have an output impedance of 75 ohms, and the antenna, if it is a normal dipole, will also have an impedance of 75 ohms, so no matching transformers will be required. It may be necessary to use a heavier-duty coax if the transmitter power is high.

You have probably noticed already that the most conspicuous difference between twin lead and coax is that one of the coax conductors is grounded, whereas neither twin-lead conductor is. Since the amplifier stages of a receiver are unbalanced—that is to say, the low side is grounded—it is necessary to use another type of transformer in the input to couple the balanced twin lead to the unbalanced amplifier stages. This transformer is called a *balun*, which is short for "balanced-to-unbalanced." This allows twin lead to be connected directly to the antenna terminals of the receiver.

Many TV receivers today come with an antenna "splitter" or "coupler." This allows more than one receiver to be connected to the transmission line, so that separate antennas are not

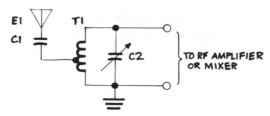

Figure A-15
Capacitive Antenna Coupling

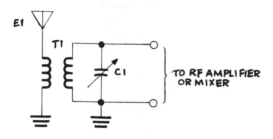

Figure A-16
Inductive Antenna Coupling

In Figure A-16, inductive coupling between the primary and secondary of T1 is used. The secondary is made resonant at a specific frequency by adjusting variable capacitor C1. At the resonant frequency the secondary impedance is high. This impedance is reflected back into the primary, so its impedance is high also at the desired frequency. Consequently, there is a greater signal voltage drop across the primary at that frequency. See *Antenna Circuits*.

Antenna Matcher (Figure A-17)

This circuit shows how a transmitter with a low-impedance output can be matched to a high-impedance transmission line. Capacitor C1 is adjusted so that the input impedance of T1 matches the output impedance of the transmitter. Capacitor C2 is tuned to the frequency of the transmitter, and the impedance of the secondary circuit is made to match that of the transmission line by selecting taps on the secondary. See *Antenna Circuits*.

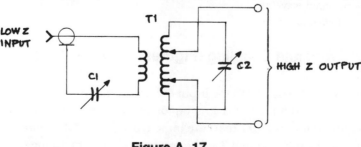

Figure A-17
Antenna Matcher

necessary. The second receiver is often an FM set. As you can se[...]
accomplished by connecting 150-ohm resistors in such a fashion th[...]
300-ohm line, and the line also sees a 300-ohm source.

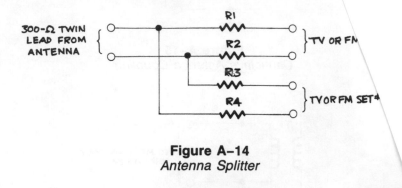

Figure A–14
Antenna Splitter

Other Antenna Circuits

See the following list for other antenna circuits described elsewhere in the encyclopedi[...]

Antenna coupling
Antenna matching
Antenna multiplexing
Antenna tuning
Balanced loop antenna
CB antenna coupler
Ferrite-core antenna
Loop antenna
Loop antenna coupling
Loop antenna, inductively coupled
Parallel-resonant wavetraps
Series-resonant wavetraps
Single series-resonant wavetrap
Transmission line connections

Antenna Coupling (Figures A–15 and A–16)

A receiving antenna can be coupled to the input of a receiver through a capacitor. As shown in Figure A–15, E1 is coupled through C1 to a tap on RF autotransformer T1. The impedance between the tap and ground is lower than the impedance across C2. If the antenna were connected through C1 to the upper end of T1, the resonant circuit consisting of T1 and C2 would be loaded down by the antenna, which would lower circuit Q, making it less selective.

Antenna Multiplexer (Figure A-18)

This circuit makes it possible for a single long-wire antenna to serve several receivers at the same time with negligible interaction. X, Y, and Z are the inputs to three receivers, each tuned to a different frequency. N represents the inputs to additional receivers. The L-C filters are tuned for series-resonance at the frequencies to which the receivers are tuned. The filters will readily pass these frequencies, but will attenuate signals at all other frequencies. See *Antenna Circuits*.

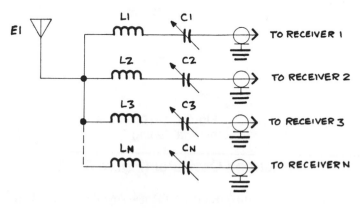

Figure A–18
Antenna Multiplexer

Antenna Splitting

See *Antenna Circuits*.

Antenna Tuning (Figure A-19)

In both figures, the received signal is developed across the primary of RF transformer T1, which is connected between the antenna E1 and ground. In (A), reducing the capacitance of the variable capacitor C1 has the same effect as reducing the length of the antenna. However, when the antenna is sufficiently long and the inductance of the primary of T1 is adequate, C1 can be adjusted to make the antenna circuit series-resonant at the receiving frequency. This results in greatly increased current in the primary, and therefore a larger voltage is developed across C2 and the secondary of T1. Selectivity is also improved when both C1 and C2 are adjusted for maximum sensitivity at the receiving frequency. C1 should have a maximum capacitance of 75 to 100 picofarads for 28 and 50 megahertz work.

If C1 is replaced by a variable inductance, as in (B), increasing the inductance of L1 has the same effect as increasing the length of the antenna. L1 can also be adjusted to make the antenna circuit series-resonant, with the same effects on selectivity and sensitivity. L1 should have an inductance around 0.1 microhenry for use at the same frequencies. See *Antenna Circuits*.

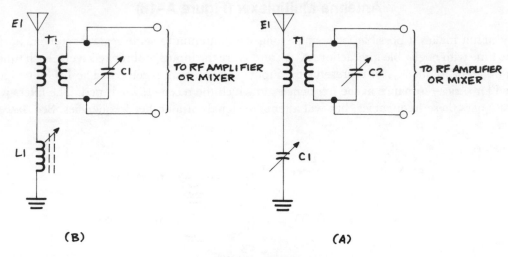

Figure A–19
Antenna Tuning

Armstrong Oscillator (Figure A-20)

In this circuit, oscillations in the tank circuit L1-C1 are sustained by positive feedback via the tickler coil L2. The parts values given below are for a frequency of approximately 3.5 megahertz. See *Oscillator and Signal Generator Circuits.*

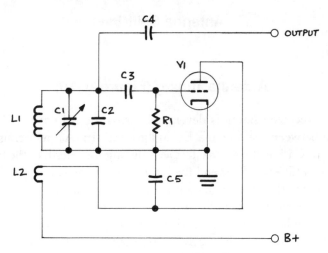

Figure A–20
Armstrong Oscillator

Parts List

C1	150 picofarads, variable
C2	500 picofarads, mica

C3, C4   100 picofarads, mica
C5   0.01 microfarad, paper
L1   4.3 microhenries
L2   Approximately one-third number of turns in L1, wound on same form, at ground end
R1   50 kilohms
V1   Low-mu triode, such as 6C5 or 6J5

Astable Multivibrator (Figure A–21)

In this circuit two transistors act as switches to turn each other on and off alternately. They are cross-connected in such a way that the collector voltage of each is applied to the base of the other. When the collector voltage of Q1 is high the base voltage of Q2 is high, and vice versa. The time interval T between pulses is given by:

$$T = 10^4 \, C \text{ (where } C = \text{ either C1 or C2)}$$

The circuit is also called a free-running multivibrator. See *Oscillator and Signal Generator Circuits.*

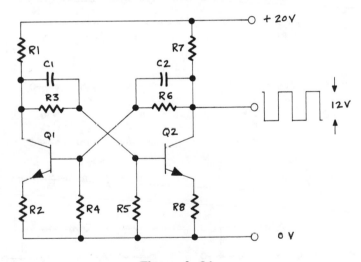

Figure A–21
Astable Multivibrator

Parts List

C1, C2   See text (depends on pulse rate desired)
Q1, Q2   2N3568 or equivalent
R1, R7   1 kilohm
R2, R8   510 ohms
R3, R6   10 kilohms
R4, R5   6.8 kilohms

ATTENUATOR AND PAD CIRCUITS

An attenuator is a network designed to provide a known reduction in the amplitude of a signal without introducing appreciable phase or frequency distortion. To do this, it has to be constructed of resistors only, and its input and output impedances must match the source and load between which it is connected.

If an attenuator's input and output impedances are the same it is said to be symmetrical, and it can be connected either way. If they are not the same it is called a minimum-loss pad. Either type may also be balanced or unbalanced. A balanced attenuator or pad has neither side grounded; an unbalanced one has one side grounded or at zero potential.

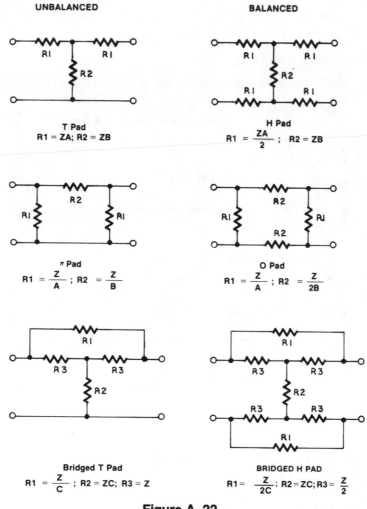

UNBALANCED

T Pad
R1 = ZA; R2 = ZB

π Pad
$R1 = \dfrac{Z}{A}$; $R2 = \dfrac{Z}{B}$

Bridged T Pad
$R1 = \dfrac{Z}{C}$; R2 = ZC; R3 = Z

BALANCED

H Pad
$R1 = \dfrac{ZA}{2}$; R2 = ZB

O Pad
$R1 = \dfrac{Z}{A}$; $R2 = \dfrac{Z}{2B}$

BRIDGED H PAD
$R1 = \dfrac{Z}{2C}$; R2 = ZC; $R3 = \dfrac{Z}{2}$

Figure A–22
Symmetrical Attenuators

Symmetrical attenuators, both balanced and unbalanced, are shown in Figure A–22. Underneath each is a formula for determining the resistance of each element. To perform the calculation, substitute for Z the required impedance, and for A, B, or C the value in the appropriate column in Table I below corresponding to the attenuation desired.

ATTENUATION dB	A	B	C
−0.1	0.00576	86.9	86.4
−0.2	0.0115	43.4	42.9
−0.4	0.0230	21.7	21.2
−0.6	0.0345	14.4	14.0
−0.8	0.0460	10.8	10.4
−1.0	0.0575	8.67	8.20
−2.0	0.115	4.30	3.86
−3.0	0.171	2.84	2.42
−4.0	0.226	2.10	1.71
−5.0	0.280	1.64	1.28
−6.0	0.332	1.34	1.00
−7.0	0.382	1.12	0.807
−8.00	0.431	0.946	0.661
−9.00	0.476	0.812	0.550
−10.0	0.519	0.703	0.462
−12.0	0.598	0.536	0.335
−14.0	0.667	0.416	0.249
−16.0	0.726	0.325	0.188
−18.0	0.776	0.256	0.144
−20.0	0.818	0.202	0.111
−22.0	0.853	0.160	0.0863
−24.0	0.881	0.127	0.0673
−26.0	0.905	0.100	0.0528
−28.0	0.923	0.0797	0.0415
−30.0	0.939	0.0633	0.0327
−35.0	0.965	0.0356	0.0181
−40.0	0.980	0.0200	0.0101
−50.0	0.994	0.00632	0.00317
−60.0	0.998	0.00200	0.00100
−80.0	0.999	0.000200	0.000100
−100.0	0.999	0.0000200	0.0000100

TABLE 1

Minimum-loss pads are shown in Figure A–23, together with formulas for calculating their resistor values. Z1 is always the larger impedance, so this part will invariably be connected to the external unit with the higher impedance.

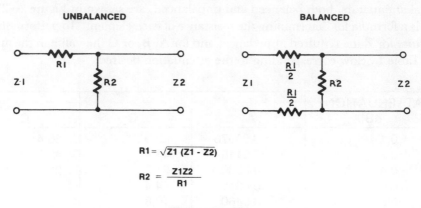

$$R1 = \sqrt{Z1\,(Z1 - Z2)}$$

$$R2 = \frac{Z1Z2}{R1}$$

Figure A–23
Minimum-Loss Pads

Attenuator Equalizer (Figure A–24)

This circuit, which is generally purchased as a complete unit, is used to correct the uneven attenuation of video signals in CCTV cables. However, it often introduces phase shift, so it must be used in conjunction with a phase equalizer. See *Passive Filters*.

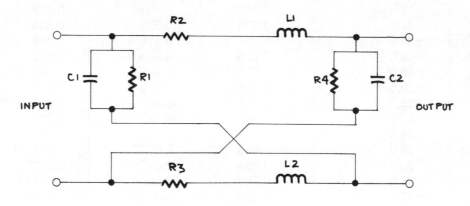

Figure A–24
Attenuator Equalizer

Audio Amplifier with Integrated Circuit
(Figure A–25)

This circuit provides 60 dB of gain over a range of 10 Hz to 16,000 Hz. The input impedance is 75,000 ohms, and maximum signal output to a 100-ohm load is 7 volts root mean square (RMS). See *Amplifier Circuits, Integrated Circuits*.

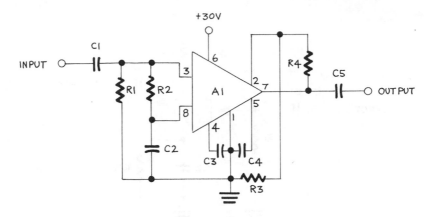

Figure A–25
Audio Amplifier with IC

Parts List

A1	Motorola MFC 8040 IC
C1, C5	1 microfarad
C2	100 microfarads
C3	0.05 microfarad
C4	0.1 microfarad
R1	75 kilohms
R2	270 kilohms
R3	100 kilohms
R4	110 kilohms

Audio Amplifier with JFET
(Figure A–26)

This circuit is similar to a triode tube. The gate is made negative with respect to the source by selecting a suitable value for the source resistor in the same way as a cathode bias resistor in a tube circuit. At the same time the bypass capacitor grounds the source for AC to avoid degeneration of signals. See *Amplifier Circuits.*

Parts List

C1, C3	20 microfarads
C2	100 microfarads
Q1	Small-signal, general purpose N-channel JFET
R1	1 megohm
R2	3.3 kilohms
R3	330 ohms

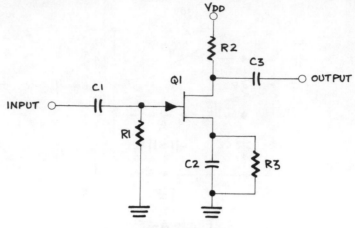

Figure A–26
Audio Amplifier with JFET

Audio Amplifier with Triode
(Figure A–27)

This is a simple audio-amplifier circuit of a type very common before the introduction of transistors. A low-mu triode tube is operated Class-A, with the grid bias set by the cathode resistor R3. B+ should be 250 volts. You can expect a voltage gain of about 15. See *Amplifier Circuits*.

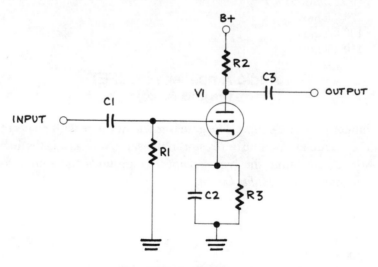

Figure A–27
Audio Amplifier with Triode

Parts List

C1	0.05 microfarad
C2	10 microfarads

C3   0.05 microfarad
R1   270 kilohms
R2   47 kilohms
R3   1500 ohms
V1   6C4 or equivalent

Audio Output Power Splitter
(Figure A-28)

(A) shows a circuit for feeding three loudspeakers from one source of audio. The sound levels of the three speakers can be controlled independently with the variable T-pads R1, R2, and R3. Transformer T1 has three 8-ohm windings. It splits the audio power so that 50 percent is available to speaker LS1, and 50 percent to speakers LS2 and LS3 combined, when LS1 is an

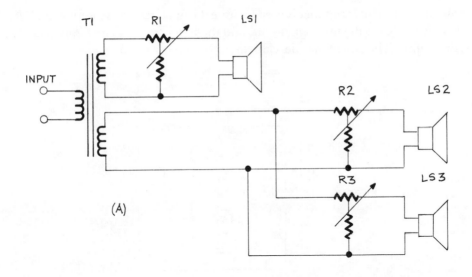

(A)

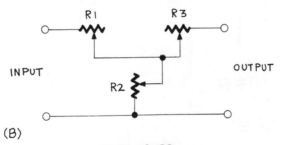

(B)

Figure A-28
Audio Output Power Splitter

8-ohm speaker, and LS2 and LS3 are 16-ohm speakers. The T-pads should have the same impedances as the speakers (R1 = 8 ohms, R2 and R3 = 16 ohms each).

(B) shows how a T-pad is constructed. The variable resistors should be ganged together to maintain the correct impedance, while varying the degree of attenuation. See *Attenuator and Pad Circuits; Amplifier Circuits.*

Parts List

R1	8-ohm T-pad
R2, R3	16-ohm T-pads
T1	Alco Mix-N-Match or equivalent

Audio Power Output Stage with IC Driver
(Figure A–29)

In this circuit an operational amplifier is used to drive the final stage of an audio amplifier. The feedback loop is via R3 to the inverting input, and the input from the previous stage is to the non-inverting input. The circuit can develop an output power of 35 watts.

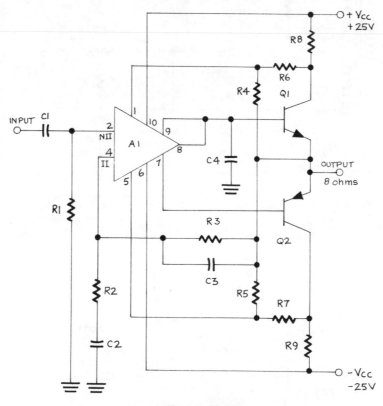

Figure A–29
Audio Power Output Stage with IC Driver

Parts List

A1	Operational amplifier, Signetics NE 540, or equivalent
C1	0.5 microfarad
C2	50 microfarads
C3	10 picofarads
C4	2000 picofarads
Q1, Q2	PNP power audio amplifier transistors with ability to dissipate at least 40 watts
R1	10 kilohms
R2	100 ohms
R3	10 kilohms
R4, R5	8.2 kilohms
R6, R7	56 ohms
R8, R9	0.18 ohm

Audio Power Output Stage with IC Driver
and Darlington Push-Pull Final
(Figure A–30)

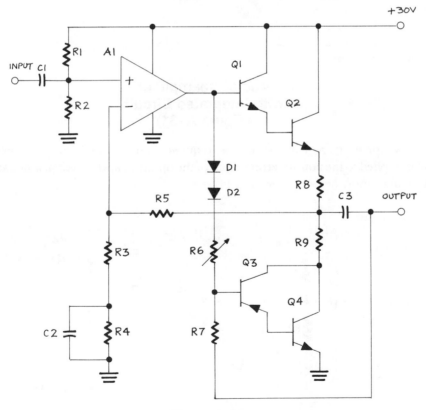

Figure A–30
*Audio Power Output Stage with IC Driver
and Darlington Push-Pull Final*

In this circuit the input signal is fed to the non-inverting input of an operational amplifier. The output signal is then applied to the two Darlington-connected transistors Q1 and Q2, and also via D1, D2 and R6 to Q3 and Q4. These four transistors form a push-pull output amplifier in spite of the fact that their input signal source is single-ended. Feedback from the junction of R8 and R9 is returned to the inverting input of A1. See *Amplifier Circuits*.

Parts List

A1	Operational amplifier, half a Fairchild µA 739, or equivalent
C1	0.1 microfarad
C2	0.5 microfarad
C3	500 microfarads
D1, D2	1N5400 series or equivalent
Q1, Q2, Q3, Q4	Power audio amplifier transistors, 3 NPN and 1 PNP
R1, R2	100 kilohms
R3	20 kilohms
R4	120 kilohms
R5	200 kilohms
R6	1 kilohm, variable
R7	15 kilohms
R8, R9	0.5 ohm

Audio Preamplifier with Integrated Circuit (Figure A–31)

This circuit will give a gain of 100 within the frequency range from 30 to 30,000 hertz. The input signal is applied to the non-inverting input of the op-amp, and a feedback signal is returned to the inverting input. See *Amplifier Circuits*.

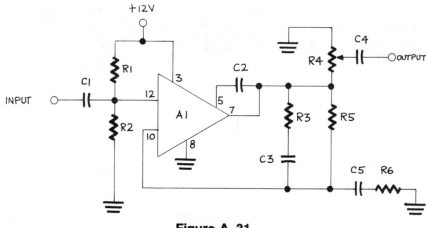

Figure A–31
Audio Preamplifier with IC

Parts List

A1	Integrated circuit, GE PA230, or equivalent
C1, C4	1 microfarad
C2	75 picofarads
C3	100 picofarads
C5	10 microfarads
R1, R2	100 kilohms
R3	10 kilohms
R4	20 kilohms, variable
R5	51 kilohms
R6	510 ohms

Automatic Gain Control
(Figure A–32)

Automatic gain control (AGC) and automatic volume control (AVC) are the same thing. The latter term is generally used in radio receivers, the former in all other equipment. Both are feedback circuits that adjust the bias voltage on a previous stage or stages to raise or lower the stage gain so as to maintain a constant output level despite changes in input signal strength. The derivation of the feedback voltage from the second detector is explained under *automatic volume control*.

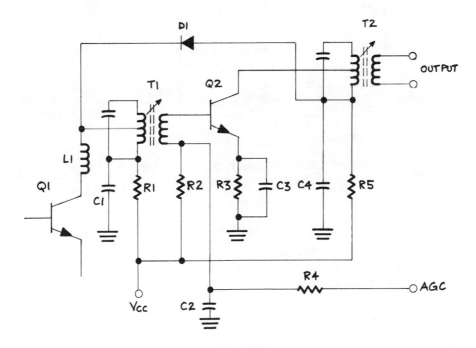

Figure A–32
Automatic Gain Control

In the circuit in Figure A–32 two methods of AGC are used together. Negative AGC voltage from the second detector is applied through R4 to the base of Q2. As the AGC voltage increases it reduces the positive forward bias on the base, so lowering the gain of the transistor. Some AGC is also provided by D1, which acts as a shunt across the primary of T1. See *Amplifier Circuits.*

Automatic Level Control
(Figure A–33)

This circuit controls the signal level by means of a light-sensitive resistance and a lamp. The input signal is fed in at phonojack J1 and out from J3. R1 and R2 form a fixed voltage divider. The light-sensitive resistive element of A1 is shunted across R2. The light-emitting element of A1 is connected across the amplifier's audio output via J2. If the audio output begins to rise, the lamp increases in brightness, which causes the resistance of the resistive element to decrease. On the other hand, if the audio output starts to fall, the decreasing brightness of the lamp causes the resistance of the resistive element to rise. In other words, A1 acts as a variable resistor in parallel with R2, causing the resistance between J3 and ground to vary, with a corresponding, but inverse, effect on the amplitude of the signal, which consequently is held level. See *Amplifier Circuits.*

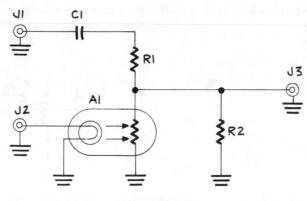

Figure A–33
Automatic Level Control

Parts List

A1	Raytheon CK1103 "Raysistor" or equivalent
C1	0.01 microfarad
J1, J2, J3	Standard phonojacks
R1, R2	2 megohms

Automatic Volume Control (Figure A-34)

In the early days of radio a common complaint was signal fading. Automatic volume control (AVC) was devised to overcome this problem by providing a means of controlling the gain of amplifier stages in inverse proportion to the strength of the received signal. The average amplitude of the signal was converted into a negative DC voltage which increased as the signal amplitude increased, and decreased as it faded. This varying negative voltage was then fed back to the grids of certain of the preceding amplifier tubes so that their gain was reduced as the signal strength increased, and vice versa.

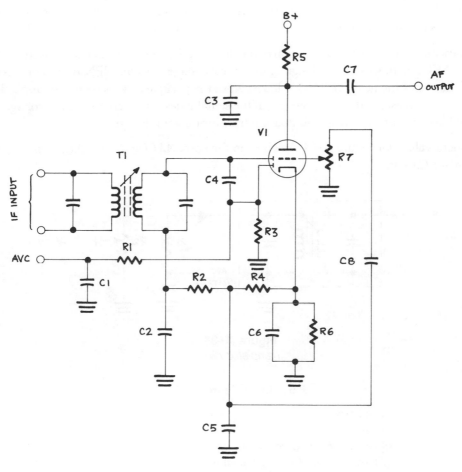

Figure A-34
Automatic Volume Control

The circuit shown in Figure A–34 is typical of standard five-tube radios of the period before the arrival of the transistor. However, transistor circuits are the same in principle, as a comparison with Figure A–35 will illustrate. In Figure A–34 the lower plate of the diode section of V1 is used for AVC rectification and is fed from the detector plate through C4. (Details of the detector action are given under *AM Detector and Audio Amplifier*.) The negative bias voltage resulting from the flow of rectified carrier current is developed across R3. This AVC voltage is applied to the grids of the controlled stages through the filter C1–R1.

In a transistor amplifier stage gain is directly proportional to base bias current within reasonable limits. The bias current may be either positive or negative, depending upon whether the transistor type is NPN or PNP. Since the AVC is designed to reduce the gain as signal strength increases, and vice versa, its polarity must be such as to oppose the bias current. This means it may be either positive or negative, according to the type of transistor used in the controlled stage.

The polarity is selected by the connection of the diode. In Figure A–35, D1 is connected to give a negative AVC, so this circuit is designed to control a stage with an NPN transistor. As signal strength increases the negative AVC voltage causes the positive base bias current applied to the transistor to decrease, and its gain lessens. If the signal fades, the decreased AVC voltage allows the base bias current to increase, so that the transistor's gain goes up.

Component values for Figure A–34 are given in the entry *AM Detector and Audio Amplifier* (q.v.). See *Amplifier Circuits*.

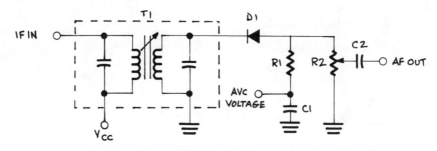

Figure A–35
Transistor AVC

Parts List (Figure A–35)

C1, C2	10 microfarads
R1	1.8 kilohms
R2	2.5 kilohms, variable (''volume control'')
T1	Final IF transformer (parts included within dashed lines are part of transformer)

AVC

See *Automatic Volume Control*.

Balanced Loop Antenna Coupling
(Figure B-1)

This is the way to connect a loop antenna that is to be used for direction finding, or any other application where an antenna connected to an unbalanced line is undesirable. The two capacitors C2 and C3 provide equal impedances to ground, while C1 tunes the antenna to resonance with the incoming signal. T1 then couples the signal to the balanced input of the receiver. No parts values are given, since they will depend upon the frequencies being used, the antenna inductance, and so on. See *Antenna Circuits*.

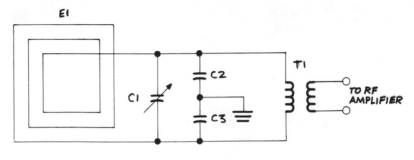

Figure B-1
Balanced Loop Antenna Coupling

Balanced Modulators
(Figures B-2 and B-3)

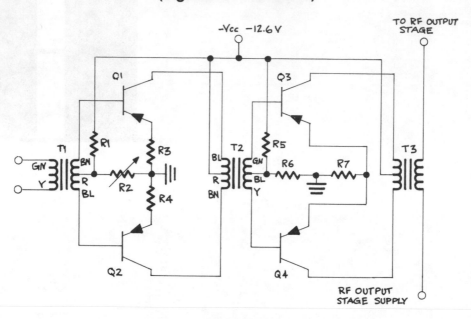

Figure B–2
Balanced Modulator (Transistor Version)

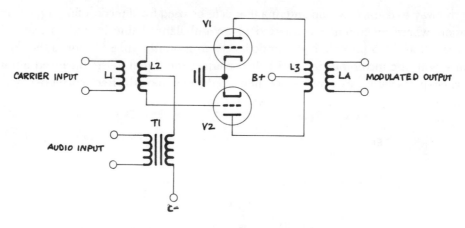

Figure B–3
Balanced Modulator (Vacuum-Tube Version)

In Figure B–2 the input from a microphone is applied across the primary of T1. A limiting resistor may be required if a carbon microphone is being used. T1 couples the audio signal to a low-power class-B push-pull amplifier Q1 and Q2. This stage is coupled in turn by T2 to a second class-B push-pull amplifier Q3 and Q4 using high-power transistors. The output signal obtained from the second stage is coupled to the transmitter's RF power amplifier by a suitable impedance-matching transformer T3. Base bias current for Q1 and Q2 is furnished through voltage divider R1 and R2, and stabilization of this stage is provided by emitter resistors R3 and R4. In the output stage, base bias current is furnished through voltage divider R5 and R6, and stabilization by the common emitter resistor R7, which also ensures balanced operation.

Figure B–3 is a vacuum-tube version of a balanced modulator. The audio signal is coupled via T1 to the grids of two triodes operated class-B in push-pull. See *Modulation Circuits*.

Parts List (Figure B–2)

Q1, Q2	2N109 or equivalent
Q3, Q4	2N278 or equivalent
R1	62 kilohms
R2	600 ohms, variable
R3, R4	47 ohms
R5	220 ohms
R6	3.3 ohms
R7	0.1 ohm
T1	Microphone transformer
T2	Interstage transformer (push-pull collectors to push-pull bases)
T3	Impedance-matching output transformer

Band-pass Active Filter (Figure B–4)

This is the circuit of a commonly-used active band-pass filter. A band-pass filter passes a band of frequencies approximately symmetrical about a center frequency f_o, with a width inversely proportional to the circuit Q, as shown in Figure B–5, where OF represents increasing frequency, and OA increasing amplitude. Freequencies f_{c1} and f_{c2} are the lower and upper cutoff frequencies at the points where the signal amplitude has decreased to 0.707 times its maximum value.

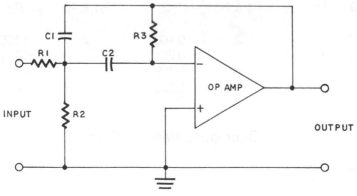

Figure B–4
Band-pass Active Filter Circuit

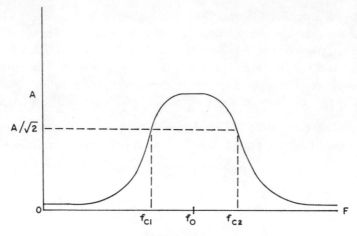

Figure B–5
Band-pass Active Filter Response Curve

Parts List

C1	Divide the required f_o in hertz into 10 to get the value in microfarads
C2	Same as value of C1 for a gain of 2; twice the value of C1 for a gain of 10
R1	15 kilohms (gain of 2); 3.3 kilohms (gain of 10)
R2	5.6 kilohms (gain of 2); 15 kilohms (gain of 10)
R3	62.8 (56 + 6.8) kilohms (gain of 2); 47 kilohms (gain of 10)

For the operational amplifier, use a 741 or 536, or equivalent.

To obtain a narrower band-pass a higher value of Q is required. This is achieved by multiplying the values for R1, R2, and R3 by the factors given in the following table:

DESIRED VALUE OF	FOR GAIN OF 2		FOR GAIN OF 10	
Q	R1 & R3	R2	R1 & R3	R2
4	2	0.400	2	0.105
6	3	0.258	3	0.061
8	4	0.189	4	0.044
10	5	0.153	5	0.034

Band-pass Passive Filter

See *Passive Filters*.

Band-reject Active Filter

See *Bandstop Active Filter.*

Band-reject Passive Filter

See *Passive Filters.*

Bandstop Active Filter
(Figure B–6)

This is the circuit of a commonly-used active bandstop filter. A bandstop filter (also called a band-reject or notch filter) rejects a band of frequencies approximately symmetrical about a center frequency f_o, with a width inversely proportional to the circuit Q, as shown in Figure B–7, where OF represents increasing frequency, and OA increasing amplitude. Frequencies f_{c1} and f_{c2} are the lower and upper cutoff frequencies at the points where the signal amplitude has decreased to 0.707 times its maximum value.

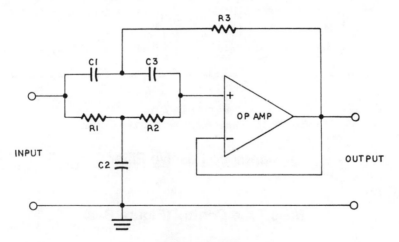

Figure B–6
Bandstop Active Filter Circuit

Parts List

C1, C3	Divide the required f_o in hertz into 10 to get the value in microfarads
C2	Twice the value of C1 or C3
R1	(Q = 1) 7.95 ohms
R2	(Q = 1) 31.8 ohms
R3	(Q = 1) 6.36 ohms

R1	(Q = 5) 1.59 ohms
R2	(Q = 5) 159 ohms
R3	(Q = 5) 1.57 ohms
R1	(Q = 10) 0.8 ohm
R2	(Q = 10) 318 ohms
R3	(Q = 10) 0.8 ohm

(These values are not available in 5 percent and 10 percent resistors, and must be specially ordered or fabricated from resistance wire.)

For the operational amplifier, use a 741 or 536, or equivalent.

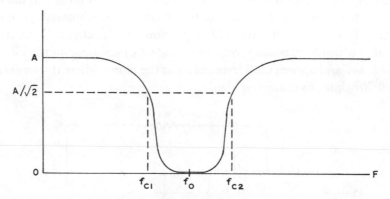

Figure B–7
Bandstop Active Filter Response Curve

Bandstop Passive Filter

See *Passive Filters*.

Bass Tone Control (Figure B–8)

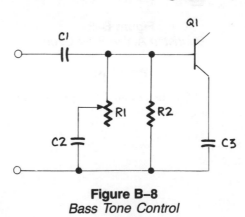

Figure B–8
Bass Tone Control

In this circuit, when the sliding contact of R1 is at the upper end of R1, C2 shunts the signal on the base of the transistor. The signals having the higher frequencies are attenuated the most, so that the bass tones are accentuated by comparison. On the other hand, if the sliding contact is at the lower end, the shunting effect of C2 is minimized, so that the bass and treble are more nearly equal. Intermediate settings of R1 give corresponding proportions of bass and treble. See *Amplifier Circuits*.

Parts List

C1, C2	0.1 microfarad
C3	100 microfarads, or according to circuit value
Q1	Transistor, according to circuit
R1	3 kilohms, variable
R2	500 kilohms, or according to circuit value

Battery Eliminator (Figure B–9)

This is the circuit of a power supply used to substitute for a battery in order to power a transistor radio or other battery-operated device. Power transformer T1 converts the line voltage to 12.6 volts AC, which is then rectified by D1. The resulting positive half-cycles charge C1. Zener diode D2 breaks down at 10.2 volts, so this potential is applied to the base of Q1 regardless of what the voltage on C1 might be. When the line voltage decreases or a load is connected, so that the output voltage tends to decrease, the forward bias on Q1 increases so that a greater current flows through the transistor. Conversely, any tendency for the output voltage to increase reduces the forward bias on Q1 and decreases the current it can pass.

With this power supply you can get a regulated output of 9 volts DC. You can get other outputs with a different T1 and D2. Substituting a 5-kilohm potentiometer for R1 and D2, with the sliding contact connected to Q1's base, will give you a variable power supply. See *Power Supply Circuits*.

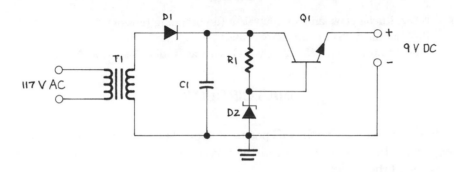

Figure B–9
Battery Eliminator

Parts List

C1   1000 microfarads, 35 volts, electrolytic
D1   Silicon rectifier diode, 6000 volts, 1 ampere
D2   Zener diode, 10.2 volts
Q1   Silicon NPN power transistor, 2N3393, or equivalent
R1   220 ohms, 1 watt
T1   Power transformer, 117/12.6 volts AC, 1 ampere

Battery Tester (Figure B–10)

When you measure the voltage of a battery with a voltmeter, you often get a "good" reading even if the battery is run down. The only satisfactory test is one that applies a load to the battery, something that a voltmeter with its high resistance cannot do. In this circuit the battery BT_x is connected to the input terminals and the pushbutton switch S1 is depressed momentarily. Current from the battery flows in the low-resistance secondary of T1, which can be a filament transformer with a 6.3-volt secondary and a 115-volt primary. When the push button is released, an inductive "kick" is induced in the primary and the neon lamp DS1 flashes if the battery is good. If there is no indication the battery no longer has enough energy to sustain a load. See *Test and Indicating Circuits.*

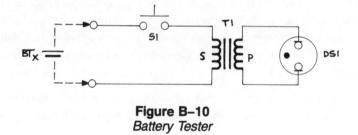

Figure B–10
Battery Tester

Parts List

DS1   NE-2H neon glow lamp or equivalent (no dropping resistor)
S1   Pushbutton, SPST, momentary contact
T1   Power transformer, 115/6.3 volts AC, such as Radio Shack Cat. No. 273–050

BIAS CIRCUITS

Bias, as understood here, means the DC potential applied to the grid of a tube with respect to the cathode, or to the base of a transistor with respect to the emitter, that establishes the operating regime of the device.

Class–A: Bias and signal voltages such that plate or collector current flows continuously throughout the electrical cycle.

Class–AB: Bias and signal voltages such that plate or collector current flows appreciably more than half but less than the entire electrical cycle.

Class–B: Bias close to cutoff such that plate or collector current flows only during approximately half of the electrical cycle.

Class–C: Bias substantially greater than cutoff, so that plate current flows for appreciably less than half of the electrical cycle.

Figure B–11(A) shows how a vacuum-tube amplifier is biased for Class–A operation. One of the things you want to do here is to obtain from the plate circuit an alternating voltage that has the same waveshape as the signal voltage applied to the grid. To do so, you must choose an operating point on the straight part of the curve. The curve must also be straight in both directions from the operating point at least far enough to accommodate the maximum value of the signal applied to the grid. If the grid signal swings the plate current back and forth over a part of the curve that is not straight, the shape of the AC wave in the plate circuit will not be the same as the shape of the grid-signal wave.

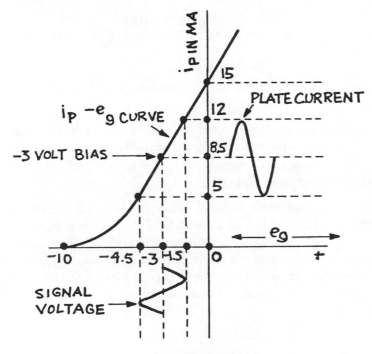

Figure B–11(A)
Vacuum-Tube Characteristic Curve

In this case the operating point is established by biasing the grid to have a potential of − 3 volts DC with respect to the cathode [Figure B–11(B)]. The most common way of doing this is by use of a cathode resistor. Since the tube has a quiescent plate current of 8.5 milliamperes, a resistor with a value of 350 ohms will provide a potential drop of 3 volts, making the cathode potential + 3 volts DC, although the cathode is still grounded for AC by its bypass capacitor. Since the

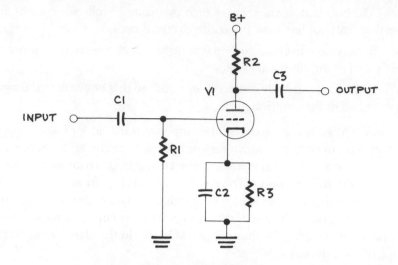

Figure B–11(B)
Vacuum-Tube Circuit for Class-A Operation

grid is at ground potential, this is the same as making it −3 volts DC with respect to the cathode.

An advantage of using a negative grid potential in vacuum-tube circuits is that the grid will not attract electrons. No signal whose peak positive voltage does not overcome the fixed negative voltage on the grid can cause grid current to flow. Consequently the tube will amplify without taking any power from the signal source.

If a transistor in a common-emitter circuit is to be operated Class–A, the operating point is established by drawing the load line on the family of characteristic curves for the device, as shown in Figure B–12. The quiescent operating point is selected at the location along the load line that gives maximum linear swing in both directions. Note that in this example the selected bias current is 40 microamperes, which gives a collector current (I_C) of 2 milliamperes and a collector voltage (E_C) of 10 volts. From this you can see that the power dissipated at the collector in the absence of a signal is $E_C \times I_C$, or $10 \times 2 = 20$ milliwatts.

Class–AB and Class–B amplifiers are push-pull circuits, in which one tube or transistor conducts while the other is cut off. They are more efficient than the Class–A amplifier, because although quiescent plate or collector current does not flow all the time (in Class–A it does), there is a greater power output.

Class-C amplifiers are used to generate RF power. The tube or tubes are delivering powerful pulses of current to a tuned circuit, and it is characteristic of a tuned circuit that it will have a high impedance at the frequency to which it is resonant, but a low impedance to all other frequencies.

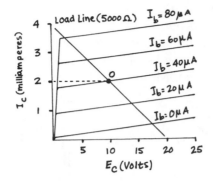

Figure B–12
Transistor Characteristic Curves

It therefore "filters out" everything but the frequency to which it is tuned, and the output is a pure sine wave at the resonant frequency. This is popularly termed "the flywheel effect." Bias is usually provided by a separate voltage source called a "C supply."

Typical Bias Circuits

See the following list for typical bias circuits described elsewhere in the encyclopedia:

AMPLIFIER CIRCUITS

Audio amplifier with triode
Broadband RF amplifier
Cathode follower
Common-emitter transistor amplifier
Linear amplifier
RF amplifier with IGFET
RF amplifier with two triodes in push-pull
Transformer-coupled push-pull power amplifier
Transistor output amplifier

Bistable Multivibrator (Figure B–13)

Bistable mutivibrators are also called flip-flops or binary counters. This is because they are the electrical equivalent of double-throw switches. Whichever of two possible states they are in, they remain in it until a second input signal flips them to the other state.

The circuit in Figure B–13 is called an Eccles-Jordan or a saturated flip-flop. It is a low-frequency (not more than 200 kilohertz) device. When power is first applied, one transistor will begin to conduct before the other (since perfect symmetry is never obtained). If Q1 starts to conduct initially, a voltage drop occurs across R1 and the negative potential on Q1's collector drops. This lowered negative potential is also felt at the base of Q2, so that its forward bias is decreased. Consequently conduction in Q2 decreases as well. This causes Q2's collector voltage to rise, since the voltage dropped across R7 is less. The higher negative voltage is also applied to the base of Q1, increasing its forward bias. As a result Q1's conduction increases further, until saturation is reached. Meanwhile Q2 becomes completely cut off.

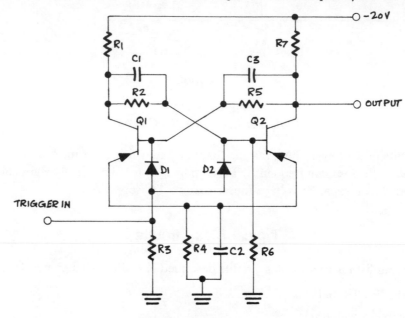

Figure B–13
Bistable Multivibrator

This stable state will remain indefinitely until a positive trigger pulse is applied at the input (across R3). When this happens, the positive voltage pulse appears on the base of Q1 (D1 is forward biased), but not on the base of Q2 (D2 is reverse biased). Q1 is immediately cut off so that current flow through R1 ceases. The collector voltage of Q1 rises to that of the power supply and is applied to the base of Q2 via R2. Q2 is now forward biased, and the current flowing through R7 results in a voltage drop that brings Q2's collector down to a low level, resulting in another stable state opposite to the first.

In the first state, the voltage on Q2's collector is high (the same as the power supply). In the second state it is low (the drop across R4 only). Consequently, a succession of positive pulses at the input will give alternate high and low voltage levels at the output. Each input pulse will be applied to the correct transistor base because the two diodes are alternately forward and reverse biased—hence they are call *steering diodes*.

An output can also be taken from the collector of Q1, so that it is perfectly possible to obtain two output signals of opposite polarity. See *Logic Circuits; Oscillator and Signal Generator Circuits*.

Parts List

C1, C2	0.002 microfarad
C3	
D1, D2	1N34A general purpose diode or equivalent
Q1, Q2	General purpose or switching PNP transistors
R1, R3	3.3 kilohms
R6, R7	
R2, R5	5 kilohms
R4	1 kilohm

Blocking Oscillator (Figure B–14)

In this circuit, when power is first applied, collector current flows through the primary of T1, building up as it overcomes the inductive reactance of the winding. During this time the changing field induces a voltage across the secondary winding of the transformer. The secondary current charges C1 with a positive polarity on the side connected to Q1's base. Since Q1 is a PNP type, this increasing positive polarity on its base reverse-biases it until it cuts off.

Now current flow stops, and C1 starts to discharge through R1. As the voltage on C1 becomes less positive, eventually changing to negative, a point is reached where Q1 starts conducting again. Now current in the primary of T1 starts building up again, inducing a voltage across its secondary, and C1 recharges as before. These cycles will continue as long as power is applied.

The time of each cycle depends mainly upon how long it takes C1 to discharge through R1, so the values of these components determine the frequency of operation. The transformer also has some influence on the frequency. Special replacement transformers for blocking oscillators are available in TV parts stores. These usually have three windings. The third winding is the output winding. This is to provide for more isolation between the oscillator and the following stage.

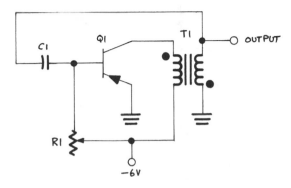

Figure B–14
Blocking Oscillator

Parts List

C1	0.15 microfarad
C2	0.10 microfarad
C3	27 microfarads, 35 volts DC
C4	100 microfarads, 20 volts DC
C5	0.15 microfarad
C6	4.7 microfarads, 6 volts DC
C7	47 microfarads, 20 volts DC
D1	General-purpose, small-signal diode
Q1, Q2, Q3, Q4	General-purpose, small-signal NPN transistors
R1	33 kilohms
R2, R14	68 kilohms
R3	820 ohms
R4	39 kilohms
R5	18 kilohms
R6, R12	680 ohms
R7	240 ohms
R8	5 kilohms, variable
R9	150 kilohms
R10	22 kilohms
R11	4.7 kilohms
R13	15 kilohms
R15	1 kilohm
Z1	Vibrating reed resonator: Ledex, or equivalent

Bridged S-meter (Figure B-17)

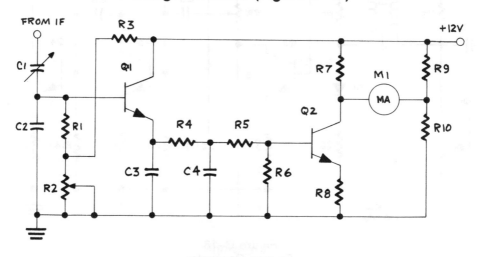

Figure B-17
Bridged S-Meter

The bridge in this circuit consists of the resistors R7 through R10, with their associated amplifier Q2. With no signal applied, R2 is adjusted so that M1 reads zero. This means, of course, that the potential on Q2's collector is the same as that at the junction of R9 and R10, so the bridge is balanced.

The input is obtained from one of the IF stages of the receiver with which this device is associated. C1 is tuned to the IF frequency, and the signal is detected by Q1. The rectified audio, after passing through the filter consisting of C3, R4, C4, and R5, is applied to the base of Q2, where it unbalances the bridge by changing Q2's collector current. The greater the degree of unbalance, the greater the reading on the meter. If the receiver is tuned to a local broadcast station to get a 10 S units indication, it is relatively simple to calibrate the dial. See *Test and Indicating Circuits*.

Parts List

C1	35 picofarads, variable
C2	10 picofarads
C3	0.01 microfarad
C4	25 microfarads
M1	0–1 milliampere DC panel meter
Q1, Q2	2N706 or equivalent
R1	33 kilohms
R2	50 kilohms, variable
R3	220 kilohms
R4	1.2 kilohms
R5	10 kilohms
R6	100 kilohms
R7, R10	4.7 kilohms
R8	470 ohms
R9	220 ohms

Bridge Rectifier (Figure B-18)

This circuit is commonly used where single-phase, full-wave rectification is required. Efficiency is good and transformer design uncomplicated. Filtering is simplified because the ripple frequency is twice the input frequency.

The operation of the circuit is easy to understand. Alternating current at the line frequency and voltage is applied to the primary of T1. An alternating current at the same frequency, but with voltage according to the turns ratio of the transformer, appears across its secondary.

When the upper end of T1's secondary is positive and its lower end is negative, electron flow is from the lower end via D4, through the load, and via D1 to the upper end. On the next half-cycle, when the upper end of T1's secondary is negative and the lower end positive, electron flow is from the upper end via D3, through the load, and via D2 to the lower end. Current flows in the same direction through the load on each half-cycle. See *Power Supply Circuits*.

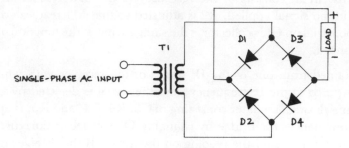

Figure B–18
Bridge Rectifier

Parts List

D1, D2, D3, D4	Individual silicon diodes or an integrated circuit may be used, as long as they have the proper ratings.
T1	This also will depend upon requirements.

Bridge Voltage Doubler (Figure B–19)

This circuit is a combination of the conventional voltage doubler (q.v.) and the bridge rectifier circuit described above. If D2 and D4 are removed, the circuit becomes a "conventional" voltage doubler. On the other hand, if C1 and C2 are removed, along with their connection to the junction of D2 and D4, the circuit becomes a standard bridge rectifier. See *Power Supply Circuits*.

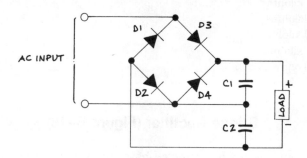

Figure B–19
Bridge Voltage Doubler

Parts List

D1, D2, D3, D4	Individual silicon diodes, or an integrated circuit may be used, as long as they have the proper ratings.
C1, C2	Not less than 16 microfarads each. The working voltage rating depends upon the output voltage of the rectifier; C2 will generally require twice the working voltage of C1.

Broadband RF Amplifier
(Figure B-20)

This circuit is an untuned, two-stage RF amplifier that uses a field-effect transistor (FET) in the input stage to provide a high input impedance and a high signal-to-noise ratio. The input signal is fed into the high impedance gate of Q1, which is connected as a common-source amplifier. The output of Q1 is applied to the base of Q2, which is connected as a common-collector amplifier.

This amplifier can handle signals from 1 to 40 megahertz. At 1 megahertz voltage gain is 40 decibels, dropping off to 26 decibels at 40 megahertz. The amplifier is designed especially for low-level signals. For instance, a 10-microvolt signal at 1 megahertz will become approximately 1 millivolt in the output.

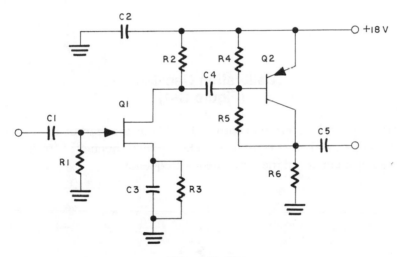

Figure B-20
Broadband RF Amplifier

Parts List

C1, C4, C5	0.003 microfarad
C2, C3	0.01 microfarad
Q1	HEP802 or equivalent
Q2	HEP51 or equivalent
R1	1 megohm
R2	470 ohms
R3	1 kilohm
R4, R5	270 kilohms
R6	4.7 kilohms

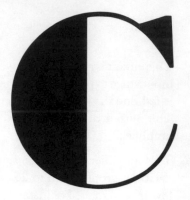

Calibration Oscillator
(Figure C-1)

This simple circuit provides a crystal-controlled oscillator that can be used for calibration of receivers, oscilloscopes, and other equipment. The crystal frequency of 100 kilohertz produces many harmonics at multiples of the fundamental frequency.

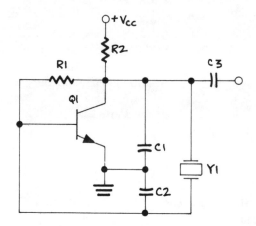

Figure C–1
Calibration Oscillator

Parts List

C1   1500 picofarads
C2   500 picofarads
C3   0.01 microfarad
D1   Any general-purpose small-signal diode
Q1   Any general-purpose small-signal NPN transistor
R1   470 kilohms
R2   3.3 kilohms
R3   1 kilohm
Y1   Crystal, 100 kilohertz

Capacitor-Input Filter (Figure C–2)

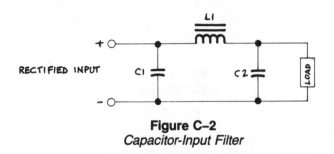

Figure C–2
Capacitor-Input Filter

This circuit is the familiar ''pi-filter'' found in so many power supplies. When used with a full-wave rectifier, the value of C1 is determined from:

$$r = 0.00188/C_1R_L \tag{1}$$

where r is the degree of filtering required and is equal to the maximum RMS ripple voltage appearing across C1, divided by the DC voltage across C1; C_1 is the capacitance of C1 in microfarads; R_L is the maximum value of the total load resistance in megohms.

The maximum value of C1 is limited, however, by the maximum allowable peak rating of the rectifier.

Additional filtering is provided by L1 and C2. The inductance L_1 in henries of L1 is obtained by dividing the maximum load reistance R_L by 1000. The capacitance C_2 of C2 is then calculated from:

$$C_2 = 100/L_1r \tag{2}$$

where r is the value found in (1) above. See *Power Supply Circuits*.

Capacitor Tester (Figure C–3)

This circuit makes a useful electrolytic capacitor tester that gives a "good-bad" indication. The capacitor to be tested is connected to the test terminals as shown. S1 is placed in the CH (charge) position, and the lamp DS1 flashes. S1 is then placed in the DIS (discharge) position, and the lamp flashes again. If the lamp does not flash it means the capacitor did not charge or discharge, which implies there is something wrong with it (assuming BT1 is not run down). A 6-volt battery is used, since many electrolytics used today have low working voltage ratings. However, you should make sure you do not use the tester on a capacitor that might be damaged. See *Test and Indicating Circuits*.

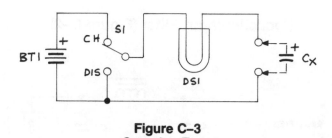

Figure C–3
Capacitor Tester

Parts List

BT1   6-volt battery, or 4 "D" cells
DS1   Type 47 incandescent lamp, complete with holder
S1   SPDT switch (any type)

Cascade Voltage Doubler (Figure C–4)

In this circuit C1 is charged to the peak value of the AC input through D2 during one half-cycle. During the other half-cycle it discharges in series with the AC source through D1 to charge C2 to twice the AC peak voltage.

The "conventional" voltage doubler (q.v.) has slightly better regulation than the cascade voltage doubler and, since the ripple frequency is twice the supply frequency, the output is easier to filter, the percentage ripple being approximately the same in both circuits. In addition, both capacitors in the conventional circuit are rated at the AC peak voltage, whereas C2 in the cascade circuit must be rated at twice this value. With both circuits, the peak inverse voltage across each rectifier is twice the AC peak.

The cascade circuit, however, has the advantage of a common input and output terminal and, therefore, permits the combination of units to give higher order multiplications (see Figure C–5). The regulation of both circuits is poor, so that only small load currents can be drawn. See *Power Supply Circuits*.

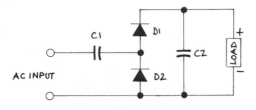

Figure C–4
Cascade Voltage Doubler

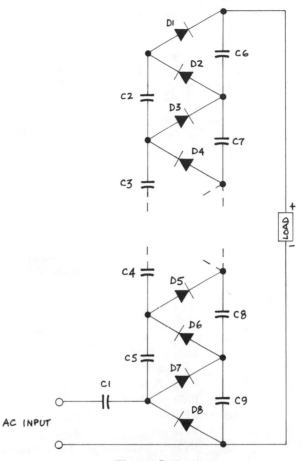

Figure C–5
Cascade Voltage Multiplier

Parts List (Figure C–4)

C1   16 microfarads (minimum), 200 working volts
C2   16 microfarads (minimum), 400 working volts
D1, D2   Silicon rectifier diodes, 1000 volts, 2.5 amperes

Cascode JFET Preamplifier
(Figure C–6)

This circuit is designed to be used between a 50-ohm antenna system and a VHF receiver with a 50-ohm input impedance. Input transistor Q1 provides about 20 decibels of gain, while Q2 reduces the impedance seen by Q1's drain. T1 and L3 form parallel resonant circuits that can be tuned by adjusting their ferrite cores. See *Amplifier Circuits.*

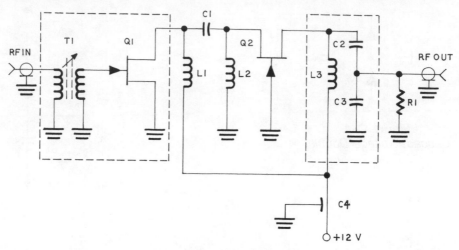

Figure C–6
Cascode JFET Preamplifier

Parts List

C1	270 picofarads
C2	5.6 picofarads
C3	20 picofarads
C4	1000 picofarads
L1, L2	RF choke, 1.2 microhenries
L3	VHF output coil, adjustable
Q1, Q2	2N5485 N-channel JFETs, or equivalent
R1	100 ohms
T1	VHF transformer, adjustable

Cascode RF Amplifier (Figure C–7)

This circuit consists of two tubes in a grounded-cathode—grounded-grid configuration. Its characteristics somewhat resemble a pentode circuit, with the advantage that no screen current is required. With the parts listed below, it is suitable for operation in the 144-megahertz band.

The grounded-grid stage V2 drastically reduces capacitive feedback from output to input, without introducing the partition noise produced by the screen of a pentode. Shot noise contributed by V2 is negligible due to the highly degenerative plate resistance of V1 in series with the cathode (as if it were a cathode resistor). The noise figure thus approaches the theoretical noise of V1 used as a triode, without the undesirable effects of triode plate-grid capacitance. See *Amplifier Circuits*.

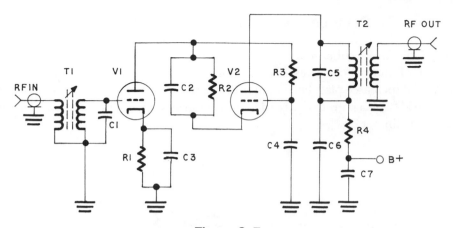

Figure C–7
Cascode RF Amplifier

Parts List

C1, C2, C3	15 picofarads, variable
C4, C5, C6, C7	470 picofarads
L1	4 turns No. 14 wire, wound into a coil 1/2 inch long, with internal diameter of 3/8 inch tapped 1-1/2 turns from "cold" end
L2	2 turns No. 14 wire in sleeving, interwound in L1
L3	3 turns No. 14 wire, wound into a coil 3/8 inch long, with internal diameter of 3/8 inch, center-tapped
L4	Same as L3, but without tap
L5	Same as L2, interwound in L4
R1, R3	220 ohms, 1/2 watt
R2, R4	470 ohms, 1/2 watt
RFC	Make RF chokes by closely winding 18 turns No. 22 enameled wire on a 1-watt carbon resistor
V1, V2	6J4 or equivalent

Cathode Bias

See *Bias Circuits*.

Cathode-Coupled Amplifier (Figure C-8)

In this circuit, two triode tubes share a common cathode resistor R2. As a result, any time that conduction in V1 increases it decreases in V2. This is because the increase in potential dropped across R2 by the increased plate current in V1 raises the voltage on V2's cathode, which is the same as if its grid (which is grounded) had become more negative. This makes the circuit useful as an amplitude limiter.

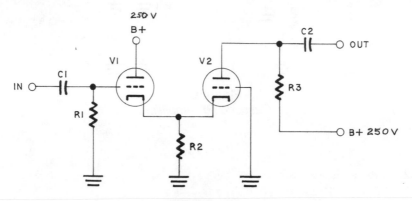

Figure C-8
Cathode-Coupled Amplifier

Parts List

C1, C2	0.1 microfarad (or according to frequency)
R1, R3	270 kilohms
R2	3.3 kilohms
V1, V2	12AT7 twin triode

Cathode Follower (Figure C-9)

This is also known as a grounded-plate amplifier, since the plate is grounded for signals through C2, thus making it the common element. The input signal is applied between the grid and plate, and the output is taken from between the cathode and plate. This circuit is degenerative, because *all* of the output voltage is fed back into the input circuit to buck the applied signal. The input signal, therefore, has to be larger than the output voltage; that is, the cathode follower not only gives no voltage gain, but actually suffers a *loss* in voltage gain. (It can still give just as much *power* gain.)

The cathode follower has two very important advantages. It has a very high input impedance (between grid and ground); and its output impedance is very low (the large amount of negative feedback has the effect of greatly reducing the plate resistance of the tube). These two

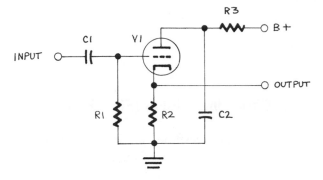

Figure C–9
Cathode-Follower Amplifier

characteristics are valuable in an amplifier that must work over a very wide range of frequencies. Also, the high input impedance and low output impedance can be used to obtain an impedance stepdown over wide ranges of frequencies that could not possibly be covered by a transformer.

No parts list is given for this cathode follower, since its design will depend very much upon the circuit in which it is used. See *Amplifier Circuits*.

CB Antenna Coupler (Figure C–10)

This circuit permits sharing a citizens band (CB) antenna by a 27-megahertz CB radio transceiver and an AM automobile radio. At the 11-meter citizens band frequencies the series-resonant combination of L1 and C1 provides a low-impedance path from the antenna to the CB transceiver. At AM broadcast band frequencies, however, the small value of C1 offers a high impedance compared to R1.

When transmitting, the outgoing signal passes easily through the low impedance of C1 and L1, but is mostly blocked from reaching the AM radio by R1, which, because of its relatively high resistance, has an insignificant loading effect on the transmitter. See *Antenna Circuits*.

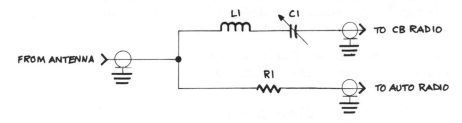

Figure C–10
CB Antenna Coupler

Parts List

C1   20 picofarads, variable
L1   2 microhenries
R1   5 kilohms

Choke-Input Filter (Figure C–11)

The percentage ripple (ripple voltage across the capacitor divided by the DC voltage across the capacitor, multiplied by 100) from a single-section filter, made up of any values of inductance and capacitance, may be determined closely enough for practical purposes for a full-wave rectifier from 100/LC, where L is in henries and C in microfarads. (In the case of a half-wave rectifier the value of LC must be doubled.) The minimum value of the choke in henries is determined by dividing the maximum load resistance in ohms by 1000, so once this is known the value of C can easily be calculated.

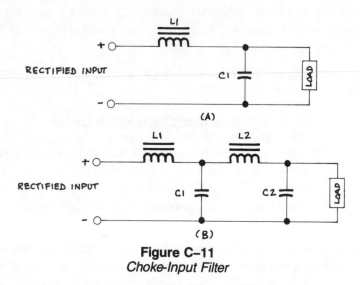

Figure C–11
Choke-Input Filter

It is obvious that the value of the choke is mathematically related to the total load resistance. Since this may vary considerably it is wise to use chokes with twice the calculated value. Alternatively, a swinging choke may be used in combination with a bleeder resistor.

Additional filter sections identical to that in the figure may be cascaded, and each will reduce the ripple at its input in accordance with the factor 100/LC given above. See *Power Supply Circuits*.

Chopper (Figure C–12)

A chopper or DC inverter is a means for changing DC to AC. There are many types, from electromechanical vibrators and dynamotors to thyratrons, ignitrons, silicon controlled rectifiers (SCRs) or thyristors, and various oscillator circuits.

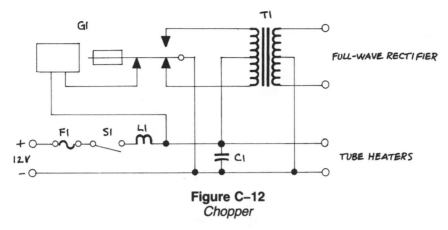

Figure C–12
Chopper

The device illustrated in Figure C–12 was standard in practically all automobile radio power supplies before the transistor replaced vacuum tubes. It was called a vibrator, an assembly consisting of an electromagnet (symbolized by the rectangle), and a reed. When S1 was closed, electron current from the negative terminal of the automobile battery flowed to the reed. Some of the current then flowed through the lower half of the primary of T1 (via the right-hand contact) and back to the positive terminal of the battery from the centertap on the winding. The rest flowed through the left-hand contact, through the electromagnet, and back to the positive terminal of the battery.

The electromagnet, activated by the current, attracted the armature on the reed. This caused it to break the connection with the two lower contacts and make connection with the upper contact. Current now flowed through the upper half of T1's primary, but in the opposite direction to the current that flowed in the lower half.

The same movement of the reed that brought this about also broke the left-hand lower contact by which the current activating the electromagnet reached it. The magnetic field therefore collapsed, and the reed returned to its former position. The initial situation was restored, and the circuit began the next cycle. The rate of vibration was between 150 and 250 cycles per second.

This chopped DC became a sine wave in T1's primary beacuse of its inductive reactance. The output from the secondary was as much as 200 volts RMS, depending upon the requirements of the B + circuit. A full-wave rectifier and pi-filter converted this AC to DC as in other power supplies.

L1 and C1 were a filter circuit to prevent sparkplug noise and other interference from the automobile engine getting into the circuit. See *Power Supply Circuits.*

Clamping (Figure C–13)

This is a simple diode circuit that clamps the collector of Q1 so that the collector-base junction does not become forward biased. If the voltage drop across R2 is kept larger than the forward drop across D1, then the base-collector junction can never become forward biased. Base current above the desired level into the base will flow through D1.

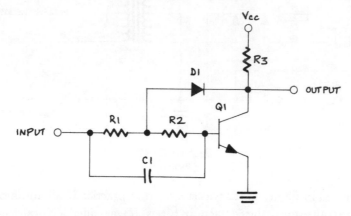

Figure C–13
Clamping to Prevent Forward-Bias

The purpose of this circuit is to ensure against saturation, since saturation slows switching time. Another method of achieving the same result is shown in Figure C–14. In this circuit, when Q1 is conducting, its collector is virtually at ground potential, because D2 is forward biased, and therefore conducting also. When Q1 turns off, its collector voltage goes negative, reverse-biasing D2, but forward-biasing D1. The potential V_L applied to D1's anode, which is less than V_{CC}, therefore appears on the collector of Q1. The output of Q1, taken from the collector, consequently is clamped at two levels, V_L and zero.

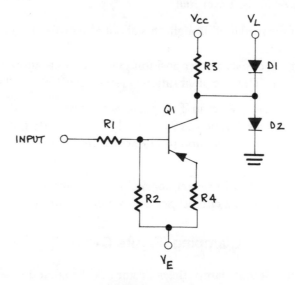

Figure C–14
Clamping a Switching Transistor's Output

Class-A Amplifier

See *Bias Circuits*.

Class-AB Amplifier

See *Bias Circuits*.

Class-B Amplifier

See *Bias Circuits*.

Class-C Amplifier

See *Bias Circuits*.

Clipper (Figure C–15)

Clippers may be either series or shunt. Figure C–15 shows a series type. The diode has been bi-ased by connecting a source of voltage, as shown. The voltage on the anode of D1 must exceed this voltage (which also appears on D1's cathode) before the diode can conduct. The diode will then pass the excess. In other words, D1 clips an input signal by the amount of the voltage source.

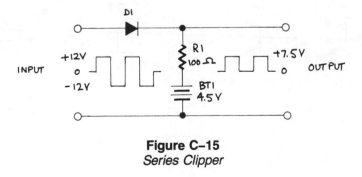

Figure C–15
Series Clipper

Figure C–16 shows a shunt clipper. When a square wave, as shown, is applied to the input, the negative excursions are not clipped because the diode is reverse biased; but the positive excur-sions forward-bias the diode, so that it conducts, and they are clipped to the level of the voltage source.

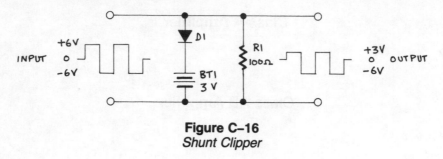

Figure C–16
Shunt Clipper

CML (ECL) NOR Gate (Figure C–17)

CML is an abbreviation for "current-mode logic," and ECL is short for "emitter-coupled logic." This type of logic achieves switching by means of nonsaturated transistors and is noted for its high speed capability.

The circuit consists of a differential amplifier (q.v.) and an emitter follower. One of the differential amplifier transistors is paralleled by an additional transistor Q1, so that either input A *or* input B will cause the differential pair to switch.

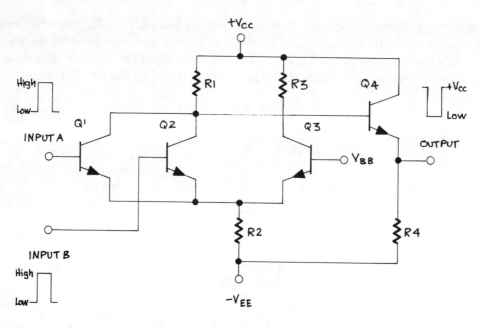

Figure C–17
CML (ECL) NOR Gate

Since Q3 has a fixed bias V_{BB}, the input voltage must exceed this before anything can happen. When it does, Q1 or Q2 turns on and Q3 turns off. Q2's collector voltage goes low because of the voltage drop across R1, and this low voltage is applied to Q4's base, turning it off. As a result, the current through R4 ceases, and the output voltage goes low. In short, a high input at either A or B, or both, results in a low output.

A complementary (OR) output can also be obtained if another emitter follower similar to Q4 is connected to get its input from the collector of Q3.

This gate is called expandable, because you can have as many inputs as you want by adding additional transistors like Q1 parallel with Q2. High fan-out operation is possible, with consequent high power dissipation. Other names for the gate are MECL (Motorola ECL) and ECCSL.

Since these devices are available in IC form, which is much less expensive and more convenient than manual fabrication, no parts list is given. See *Logic Circuits*.

CML OR Gate

See *CML (ECL) NOR Gate* (complementary output).

Code-Practice Oscillator (Figure C-18)

This is a simple oscillator that can be used for code practice by interrupting the supply voltage with a telegraph key. Basically an amplifier, the circuit is caused to oscillate by means of positive feedback to the base of Q1 via C1. Tone control is by means of R1, which varies the amount of bias on the base of Q1. Changing the value of C1 will also vary the frequency of oscillation. See *Oscillator and Signal-Generator Circuits*.

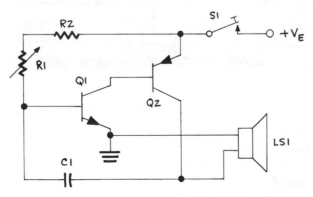

Figure C-18
Code Practice Oscillator

Parts List

C1	15 microfarads
LS1	Any small speaker
Q1, Q2	General-purpose small-signal NPN transistors
R1	250 kilohms, variable
R2	22 kilohms

Colpitts Oscillator, Crystal Controlled
(Figure C–19)

This is a practical, crystal-controlled Colpitts oscillator suitable for use in transmitters, as a test oscillator, or other applications. See *Oscillator and Signal-Generator Circuits.*

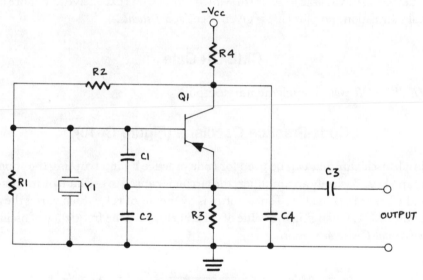

Figure C–19
Colpitts Oscillator (Crystal Controlled)

Parts List

C1	330 picofarads
C2	68 picofarads
C3	150 picofarads
C4	0.01 microfarad
Q1	General-purpose small-signal PNP transistor
R1	4.7 kilohms
R2	12 kilohms

R3   3.3 kilohms
R4   1 kilohm
Y1   Crystal, 100 kilohertz fundamental frequency

Colpitts Oscillator, Series Fed
(Figure C–20)

In this circuit, collector supply voltage is fed through L1. This inductor, in conjunction with C1 and C2, determines the frequency of oscillation. See *Oscillator and Signal-Generator Circuits*.

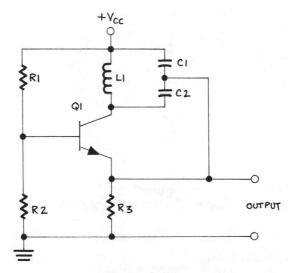

Figure C–20
Colpitts Oscillator (Series Fed)

Parts List

C1, C2   Selected for desired frequency f from following approximate formula
L1*   Selected for desired frequency f from following approximate formula

$$f^2 = (C1 + C2) / L1(C1C2)$$

where C1 and C2 are in farads and L1 in henries

Q1   General-purpose small-signal NPN transistor
R1   121 kilohms
R2   8.2 kilohms
R3   1.5 kilohms

*L1 should have an adjustable ferrite core for fine tuning.

Colpitts Oscillator, Shunt Fed
(Figure C–21)

This circuit is shown using a vacuum tube. It is designed for a frequency of 3.5 megahertz, with fine tuning provided by C1. See *Oscillator and Signal-Generator Circuits.*

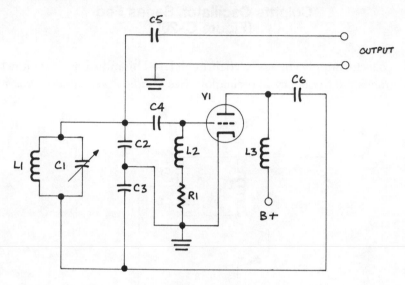

Figure C–21
Colpitts Oscillator (Shunt Fed)

Parts List

C1	150 picofarads, variable
C2	700 picofarads, mica
C3	0.0021 microfarad, mica
C4	100 picofarads, mica
C5	100 picofarads, mica
C6	0.001 microfarad, mica
L1	4.3 microhenries
L2, L3	RF chokes, 2.5 millihenries
R1	50 kilohms
V1	6C5 or 6J5, or equivalent

Common-Base Transistor Amplifier
(Figure C–22)

This is one of the three basic transistor amplifier circuits. Each is identified by the transistor element that is common to both the input and the output. Very often this common element is connected to the circuit's ground. The other configurations are the common-emitter and common-collector circuits (q.v.).

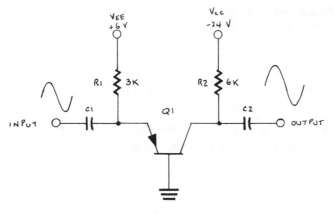

Figure C–22
Common-Base Transistor Amplifier

The common-base amplifier is often used as a low-to-high impedance coupling circuit, which is frequently needed in the input of a receiver between the transmission line and the first RF amplifier. However, in analyzing this circuit, you can consider it in general terms to be "on its own."

Figure C–23 gives you a set of characteristic curves for the transistor. Start by drawing the load line. Since E_c (V_{CC}) is – 24 volts, mark this on the horizontal axis (which represents power supply voltage).

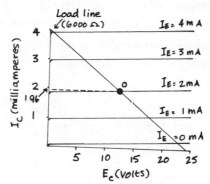

Figure C–23
Transistor Characteristic Curves

The maximum current flowing in R2 is given by – 24 / 6000 = – 4 milliamperes, so mark this on the vertical axis of the graph. The line joining this point and the point on the horizontal axis is the load line.

You now have to select the operating point for the transistor. This must be at about the midpoint of the load line so that both positive and negative voltage swings of the output signal will be on that part of the load line that intersects with linear portions of the characteristic curves.

This is where the bias current into the emitter I_E is 2 milliamperes. Thus, the bias resistor R1 is equal to $6 / 2 \times 10^{-3} = 3000$ ohms. (The emitter resistance itself is usually less than 50 ohms, so it can be neglected.)

This transistor has an alpha of 0.98; therefore, I_C is about 2 percent less than I_E, or 1.96 milliamperes. This produces a voltage drop across R2 of 11.76 volts and leaves 12.24 volts from collector to ground.

Since the current gain is equal to alpha it is less than 1. However, the voltage gain G_V is obtained by multiplying alpha by the ratio of the output resistance to the input resistance:

$$G_V = 0.98 \times 6000 / 12.8 = 459$$

That mysterious 12.8 ohms is a theoretical value for the emitter resistance that is roughly correct for most transistors. It is the principal fly in the ointment, since there are not too many signal sources that match or are lower than this figure. Consequently it is not easy to take full advantage of this high voltage gain. It should also be noted that in a common-base circuit the signal is not inverted at the output. See *Amplifier Circuits*.

Common-Collector Transistor Amplifier
(Figure C–24)

This amplifier has exactly the opposite characteristics to the common-base amplifier (q.v.). It has high current gain, less-than-unity voltage gain, and high input resistance. In these ways it resembles the vacuum-tube cathode follower, so it is often called an emitter follower.

Assuming this transistor has an alpha of 0.98, the input resistance is given by the following approximate equation, in which $R_o = R3 = 1500$ ohms:

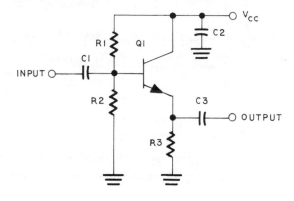

Figure C–24
Common-Collector Transistor Amplifier

$$R_i = R_o / (1 - alpha)$$
$$= 1500 / (1 - 0.98)$$
$$= 75000 \text{ ohms}$$

In this circuit, the voltage gain is equal to alpha. That is to say, for this transistor it is 0.98, or roughly 1. The approximate value for current amplification A_i or power gain G is given by:

$$A_i = (1 - alpha)^{-1} = G$$
$$= (1 - 0.98)^{-1} = G$$
$$= 50 = G$$

See *Amplifier Circuits*.

Common-Emitter Transistor Amplifer

See *Amplifier Circuits*.

Compression Amplifier
(Figure C–25)

The function of a compression amplifier is to stabilize the level of the audio signal fed to the input of a sound system or radio transmitter. This is necessary when the performer is moving about, causing the microphone output level to vary. In this circuit, automatic gain control (q.v.) is used to reduce gain when input level rises, and vice versa.

The microphone is connected to J1, and its signal is applied via C2 to the control grid of V1. After amplification it is fed to the grid of the cathode follower V2A. The desired output level is set by adjustment of R10 and appears at J2.

The output signal from V1 is also applied to the grid of V2B, where it receives further amplification, and then is coupled via T1 to the voltage-doubler rectifier consisting of D1, D2, C10, and C11. The resulting DC AGC voltage is applied via R2 to the third, or injection grid of V1. Since this is a negative voltage, an increase in it causes V1's gain to decrease; conversely, a decrease has the opposite effect. Rapidity of response depends on the time constant of C1 and R2. See *Amplifier Circuits*.

Parts List

C1	1 microfarad
C2, C8	0.01 microfarad
C3, C4, C7, C9	4 microfarads
C5	0.05 microfarad
C6	0.02 microfarad

C10, C11	0.2 microfarad
D1, D2	1N34A or equivalent
J1, J2	C3M-type 3-pin chassis-mount audio connector (third pin may be used for push-to-talk function, if used)
R1, R9	220 kilohms
R2, R4	100 kilohms
R3	220 ohms
R5	2.2 kilohms
R6, R7, R8	470 kilohms
R10	2.5 kilohms, variable
R11	1.5 kilohms
R12	1 kilohm
R13	47 kilohms
R14	12 kilohms
T1	Interstage coupling transformer
V1	12BE6 pentagrid
V2A, V2B	Two halves of a 12AU7A twin diode

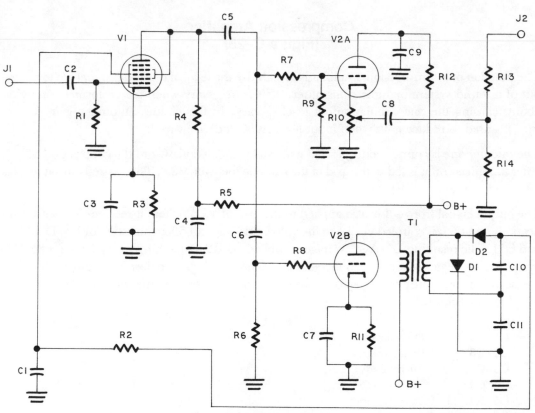

Figure C–25
Compression Amplifier

Constant-Gain Audio Amplifier
(Figure C–26)

This amplifier is intended for use in low signal level applications where automatic adjustment of the signal level is desired. The input signal is coupled by T1 to the grids of the push-pull amplifier consisting of V1 and V2. The output signals from this amplifier are applied to the grids of V4 and V5, which also operate in push-pull. The output signals from V4 and V5 are combined in T2 and coupled to the next stage.

The gain of V4 and V5 is held constant by R14. V1 and V2 do not have a cathode resistor; instead, they have a variable bias circuit in which V3's conduction varies in accordance with a feedback signal from V5. As the output signal on the plate of V5 increases, the signal on V3's grid also increases, so that it conducts more heavily. This lowers its plate voltage, and consequently makes the grids of V1 and V2 more negative, reducing their gain. (The plate of V3 is negative, but its cathode is more negative.) When the signal on V5's plate decreases, the opposite result is obtained. See *Amplifier Circuits*.

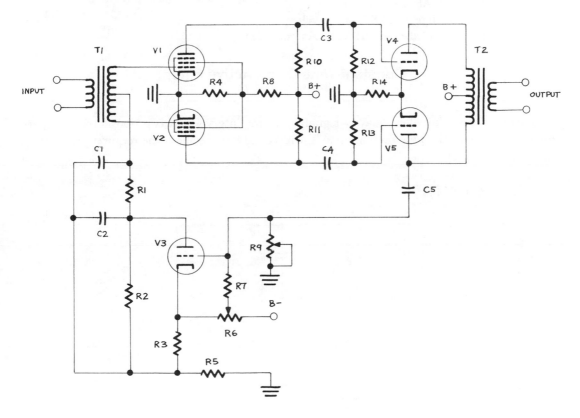

Figure C–26
Constant-Gain Audio Amplifier

Parts List

C1, C2	2 microfarads
C3, C4	0.05 microfarad
C5	0.01 microfarad
R1	10 megohms
R2, R7, R12, R13	470 kilohms
R3	3.9 kilohms
R4	5.9 kilohms
R5	200 ohms
R6	2.5 kilohms, variable
R8	30 kilohms
R9	1 megohm, variable
R10, R11	100 kilohms
R14	1.5 kilohms
T1, T2	Interstage coupling transformers, with centertaps
V1, V2	12BA6 or equivalent
V3	Half of a 12AX7, or equivalent
V4, V5	Two halves of a 12AU7A, or equivalent

Constant-Time-Delay Active Filter
(Figure C–27)

A constant-time-delay filter delays a signal by a time interval that is proportional to the frequency of the signal. If the time interval T_o (in seconds) is specified, the required frequency f_o (in hertz) is given by:

$$f_o = 6 / 40.84 T_o$$

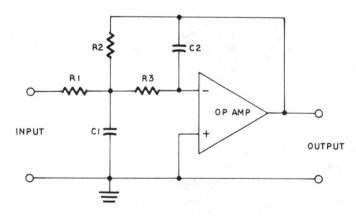

Figure C–27
Constant-Time-Delay Active Filter

Conversely, the time delay T_o (in seconds) for a certain frequency f_o is given by:

$$T_o = 6 / 40.84 f_o$$

The value of C1 is found by dividing the required f_o into 10, which gives C1's capacitance in microfarads. Values of other components are given in the following table, where capacitances are in microfarads and resistances in kilohms:

COMPONENT	FOR GAIN OF 2	FOR GAIN OF 10
C2	0.200 × C1	0.047 × C1
R1	10.99	7.49
R2	21.97	74.90
R3	19.18	23.95

See *Active Filter Circuits*.

Constant-Voltage Speaker Feed
(Figure C–28)

In a system like this, the sound level of each speaker can be set without affecting the level of the other speakers, even though they have different impedances or wattages. By using transformers with multiple taps on their secondary windings, the audio signal on the line (which may be 115, 70, or 25 volts) can be adapted to each speaker's requirements.

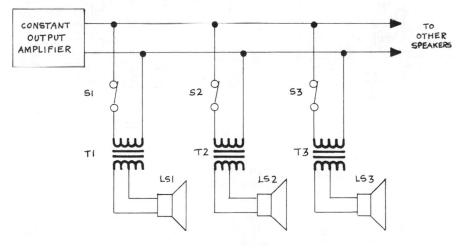

Figure C–28
Constant-Voltage Speaker Feed

For instance, if LS1 has an impedance of 8 ohms, and should have a power input of 12.5 watts, then since $E^2 = PZ$ (neglecting phase shift, if any), the voltage output from the secondary of T1 should be equal to the square root of 12.5×8, or 10. A suitable tap should be selected to give this voltage. Similarly, if LS2 has an impedance of 4 ohms, and should have a power input of 1 watt, T2 should be tapped for an output of 2 volts RMS.

Converter (Figure C–29)

This typical circuit employs a single transistor in a common-emitter circuit. In operation, RF signals are picked up by antenna coil L1, which is tuned by variable capacitor C1A. A step-down secondary winding L2 on the ferrite core matches the high impedance of the tuned circuit to the moderate input impedance of the transistor. This assures efficient transfer of signal energy. It also minimizes the loading on the tuned circuit, so that its selectivity is not impaired. The signal from L2 is coupled by C2 to the base of Q1. Q1's base bias is fixed by the voltage divider R1 and R2, in conjunction with the emitter resistor R3.

Q1's collector is connected to L4, which feeds back energy to L3; the energy is coupled via C4 to the emitter to start and maintain oscillation. The tap on L3 minimizes the loading effect of R3, ensuring good circuit Q. This coil is tuned by C1B, which is ganged with C1A. C3 is a series padder and C5 a trimmer capacitor.

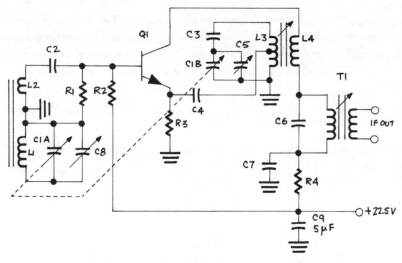

Figure C–29
Transistor Converter

There are four frequencies present in the collector circuit of Q1: the received signal, the oscillator signal, and the sum and difference signals resulting from the RF and oscillator signals beating together. One of the latter is selected by the IF transformer T1 and coupled to the first

IF amplifier. R4 and C7 form a decoupling filter in the collector supply circuit. See *Demodulation Circuits*.

Parts List

C1	Two-section variable tuning capacitor: C1A, 4.8–74 picofarads; C1B, 6.2–59 picofarads
C2	0.02 microfarad
C3	268 picofarads
C4	0.01 microfarad
C5, C8	2–14 picofarads, variable
C6	100 picofarads
C7	0.001 microfarad
L1, L2	Ferrite loop antenna
L3, L4	Oscillator coil
Q1	2N168A or equivalent
R1	82 kilohms
R2	56 kilohms
R3	18 kilohms
R4	2.2 kilohms
T1	IF transformer

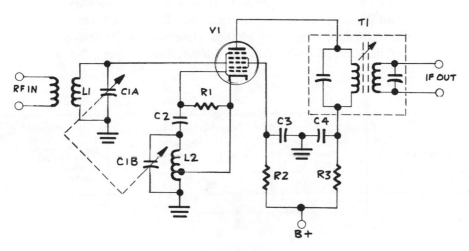

Figure C–30
Pentagrid Converter

Figure C–30 shows a pentagrid converter. Grids 1 and 2 and the cathode are connected to a conventional oscillator circuit and act as a triode oscillator. Grid 1 is used as the grid of the oscillator and grid 2 acts as the plate. In this circuit the first two grids and the cathode can be considered as a composite cathode, which supplies to the rest of the tube an electron stream that varies at the oscillator frequency. The signal voltage is applied to grid 3, which controls the electron stream so that the plate-current variations are a combination of the oscillator and the incoming signal frequencies, from which the IF transformer selects the desired frequency. See *Demodulation Circuits*.

Parts List

C1	Two-section tuning capacitor similar to C1 in Figure C–29, 40–350 picofarads
C2	47 picofarads
C3, C4	0.001 microfarad
L1	240 microhenries
L2	130 microhenries
R1, R2	22 kilohms
R3	1 kilohm
T1	IF transformer
V1	12BE6

Cowan Bridge Balanced Modulator
(Figure C–31)

The carrier is applied across C and D. Without any modulating signal the bridge is balanced, and there is no output (the carrier is suppressed). However, when a modulating (audio) signal appears across A and B, the diodes are switched on and off at the audio rate, unbalancing the bridge. This results in the production of sidebands, generated by the heterodyning of the two signals. The output, therefore, consists of double sidebands with suppressed carrier. See *Modulation Circuits*.

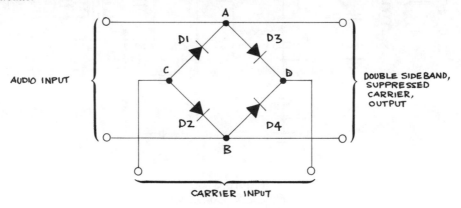

Figure C–31
Cowan Bridge Balanced Modulator

Parts List

D1, D2, D3, D4	1N914/4148 silicon switching diodes, or equivalent, with adequate rating

Crystal-Controlled Oscillator

See specific oscillator, or *Oscillator and Signal Generator Circuits.*

Crystal Detector (Figure C–32)

The simplest radio receiver consists of a diode, headphones, antenna, and ground. In the early days of radio, when there was only one local station anyway, the diode was a galena (native lead sulfide) crystal. A point contact between a conducting metal and a cleavage surface made an efficient rectifier for small RF signals. The contact consisted of a springy wire "catwhisker" which was moved from place to place until a "sensitive" spot was found. This was called "tickling the crystal." Early receivers were, therefore, called "crystal sets." Today, a semiconductor diode would be used, and some way of tuning the desired frequency (so as to eliminate the unwanted stations) would be required. See *Demodulation Circuits.*

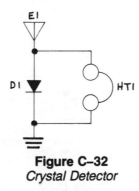

Figure C–32
Crystal Detector

Parts List

D1   1N34A germanium diode
HT1   Headphones, at least 20,000 ohms impedance

Crystal Lattice Filter (Figure C–33)

In this circuit, four crystals form a lattice filter to pass one of the sidebands of a modulated RF signal. This sideband extends from 9.000 megahertz (the carrier frequency) to 9.002 megahertz. Y1 and Y2 are resonant at 9.0005 megahertz, and Y3 and Y4 at 9.0015 megahertz. See *Passive Filters.*

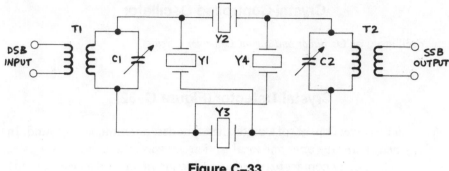

Figure C–33
Crystal Lattice Filter

Crystal Set (Figure C-34)

The crystal detector (q.v.) cannot tune stations, but this is provided for in the crystal set circuit by addition of L1 and C1. C2 is to bypass RF. See *Demodulation Circuits.*

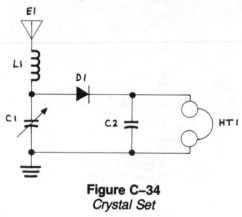

Figure C–34
Crystal Set

Parts List

- C1   Single-section tuning capacitor 30–350 picofarads
- C2   0.01 microfarad
- D1   1N34A germanium diode or equivalent
- L1   108 turns of No. 22 enameled copper wire wound on a 1-inch coil form (about 3 inches long)

CTL OR Gate (Figure C-35)

CTL stands for "complementary transistor logic." Like CML, it is a high-speed gate, but because of the more difficult manufacturing process it is more expensive. When either or both Q1 and Q2 receive a negative voltage, they turn on, and the $-V_{cc}$ potential is connected to the base of Q3. Since this is an NPN transistor, it is turned off, so that the $-V_{CC}$ potential appears on its emitter also. Since this gate is supplied in IC form, no parts list is given here. See *Logic Circuits*.

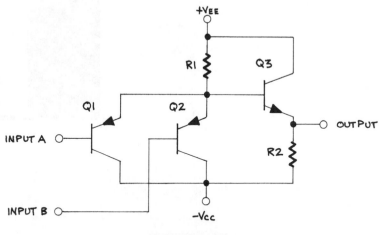

Figure C–35
CTL OR Gate

Current and Voltage Feedback Amplifier
(Figure C–36)

This circuit uses both current and voltage feedback to compensate for the fact that output impedance is increased by current feedback and decreased by voltage feedback. Therefore a combination of the two, as shown, may be used to provide constant output impedance when required. Voltage feedback is obtained from the junction of R4 and R5. Current feedback is developed across R2 when it is not bypassed at the signal frequencies. Parts values will depend on the circuit in which it is used. See *Amplifier Circuits*.

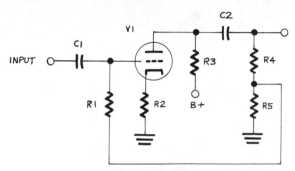

Figure C–36
Current and Voltage Feedback Amplifier

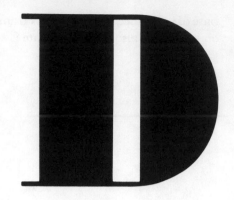

Darlington Amplifier (Figure D-1)

In this circuit, Q1 drives Q2 directly from emitter to base. This arrangement provides high current gain but unity voltage gain. It makes a very useful power output circuit for driving low impedance loads such as loudspeakers. Combining it with a second similar circuit of opposite polarity creates a complementary symmetry power output circuit which is in wide use. A good

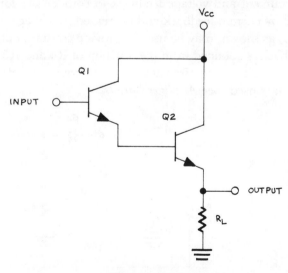

Figure D-1
Darlington Amplifier

example of this use is given in the entry *Audio Power Output Stage with IC Driver and Darlington Push-Pull Final*. See also *Amplifier Circuits*.

Darlington Differential Amplifier
(Figure D–2)

This is the schematic diagram of an integrated circuit, so there is no point in giving parts values. Q1 and Q2 are a Darlington pair, as are Q4 and Q5. Q3 replaces the common emitter resistor used in a differential amplifier, and its resistance is held constant by the fixed bias voltage V_{BB}. For an explanation of how a differential amplifier works, see the entry *Differential Amplifier*. See also *Amplifier Circuits*.

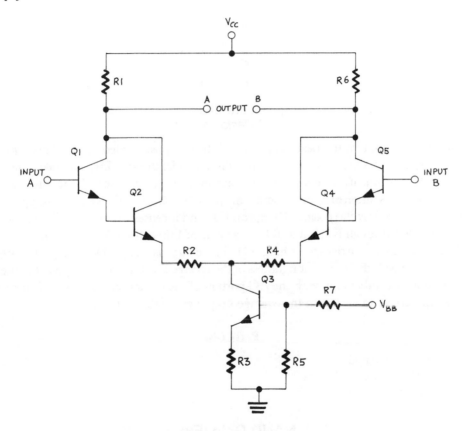

Figure D–2
Darlington Differential Amplifier

DC Restoration (Figure D–3)

In a video amplifier, coupling between stages via a capacitor will eliminate the brightness level, because it is a DC component of the signal. It is necessary to restore this if the picture is to have proper contrast, and if the sync pulse tips are to be realigned evenly.

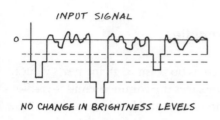

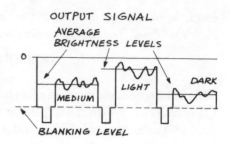

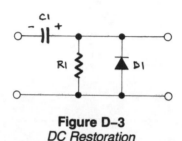

Figure D–3
DC Restoration

In this circuit, suppose that the peak amplitude of the composite video signal applied to the DC restorer starts out at 50 volts, so C1 charges up to this potential through D1. This takes place almost instantaneously, because of the very short time constant presented by the capacitor and the conducting diode. At the conclusion of each sync pulse the positive voltage of D1's anode disappears, so the diode is reverse biased by the positive charge on C1, and it ceases to conduct. The charge on C1 begins to leak away through R1, but as R1 has a value of 1 megohm it loses very little of its charge before the next sync pulse arrives to recharge C1. This charge is really a DC voltage, the amplitude of which will vary with the average brightness of the received picture. When applied to the grid of the picture tube in combination with the AC portion of the composite video signal, it causes the average beam current to vary with the average brightness of the picture.

Parts List

C1   0.05 microfarad
D1   1N4000 series, 200 volts
R1   1 megohm

DCTL NAND Gate (Figure D–4)

DCTL is short for "direct-coupled transistor logic." In this circuit, with a logical 0 on the bases of Q1 and Q2, both transistors are cut off. Consequently, the output is the same as the V_{cc} supply voltage and is a logical 1. When a logical 1 is applied to both inputs, Q1 and Q2 both turn on, and practically all of V_{cc} is dropped across R1. Now the output is close to zero, or a logical 0. See *Logic Circuits*.

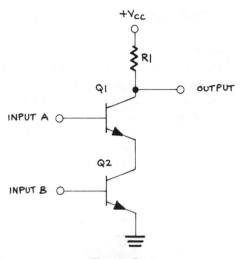

Figure D–4
DCTL NAND Gate

Parts List

Q1, Q2	Any available switching transistors
R1	100 ohms (or according to I_C)

DCTL NOR Gate (Figure D–5)

In this circuit, Q1 and Q2 are parallel, so either of them can, by turning on, cause the output to go to zero (logical 0). Parts values are the same as for the DCTL NAND gate. See *Logic Circuits*.

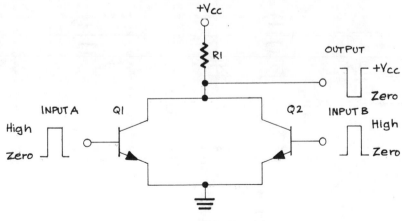

Figure D–5
DCTL NOR Gate

De-Emphasis (Figure D-6)

FM broadcast transmitters pass the signal through a pre-emphasis network (q.v.) to emphasize the amplitude of the higher frequencies with respect to the lower. In a receiver, therefore, a de-emphasis circuit is required to restore the frequencies to their proper proportions. The circuit consists of a capacitor and resistor with values such that their RC constant is 75 microseconds. See *Demodulation Circuits; Passive Filter Circuits*.

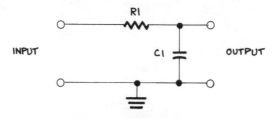

Figure D-6
De-Emphasis

Degenerative Neutralization
(Figure D-7)

In this circuit V1 is an RF amplifier and V2 is a grid leak detector. If both the grid and plate circuits of V1 are tuned to the same frequency, the stage will oscillate if steps are not taken to prevent it.

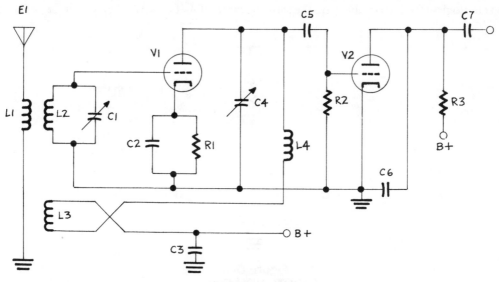

Figure D-7
Degenerative Neutralization

In this case, oscillation is prevented by making the plate load of V1 consist of two coils L3 and L4 connected in series. The combined inductance of these coils is such that the plate circuit can be tuned over its intended frequency range by C4. L3 is inductively coupled to L2, but with opposite polarity, so that the feedback is degenerative (negative). The amount of feedback can be controlled by varying the coupling between L3 and L2. See *Amplifier Circuits*.

Parts List

C1, C4	Values depend on frequencies to be tuned
C2	680 picofarads
C3	470 picofarads
C5	100 picofarads
C6	680 picofarads
C7	0.1 microfarad
L1, L2, L3, L4	Values depend on frequencies to be tuned
R1	1.5 kilohms
R2	1.5 megohms
R3	100 kilohms
V1, V2	Two halves of a 12AX7

Delayed AVC

In Figure A–31, the audio-diode return is made directly to the cathode and the AVC-diode return to ground. This places negative bias on the AVC diode equal to the DC drop across the cathode resistor (a volt or two), and thus delays the application of AVC voltage to the amplifier grids, since no rectification can take place in the AVC-diode circuit until the carrier amplitude is large enough to overcome the bias. Without this delay the AVC would start working even with a very small signal. This is undesirable, because the full amplification of the receiver then could not be realized on weak signals. See *Demodulation Circuits*.

DEMODULATION CIRCUITS

Demodulation, also called detection, is the process of recovering the modulation from a signal. Any device that is nonlinear (whose output is not *exactly* proportional to its input) will act as a detector. The author once possessed a demodulating bathtub. Located only about a mile from KFI, a Los Angeles broadcasting station with a 50-kilowatt output, it would demodulate the station signal by means of some nonlinear element in the plumbing, and the resulting audio was sufficiently powerful to vibrate the tub enough to be heard plainly by anyone taking a bath!

Whatever the device is, it can be used as a detector if an impedance for the desired modulation frequency is connected in the output circuit, so that the detector output can develop across this impedance.

Detector sensitivity is the ratio of desired detector output to the input. Detector linearity is a measure of the ability of the detector to reproduce the exact form of the modulation on the incoming signal. The resistance or impedance of the detector is the resistance or impedance it presents to the circuits it is connected to. The simplest AM detector is a semiconductor rectifier, usually consisting of a silicon or germanium diode, as in Figure D–8. The progress of the signal through this circuit is illustrated in Figure D–9. A typical modulated signal as it exists at the RF input is shown at (A). When this signal is applied to the diode, current can flow only during the part of each RF cycle when the anode is positive with respect to the cathode, so that the output of the rectifier consists of half-cycles of RF still modulated as in the original signal, as in (B).

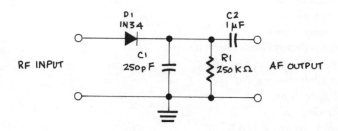

Figure D–8
Diode Detector Circuit

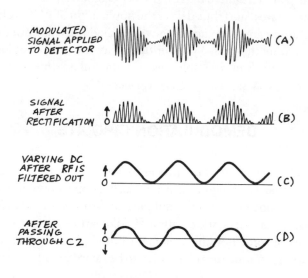

Figure D–9
Diode Detector Theory of Operation

These current pulses flow in the load circuit comprised of C1 and R1. The values of C1 and R1 are so proportioned that C1 charges to the peak value of the rectified voltage on each pulse, and retains enough charge between pulses so that the voltage across R1 is smoothed out, as shown in (C). C1 thus acts as a filter for the radio frequency, leaving a DC component that varies in the same way as the modulation on the original signal. When this varying DC is applied to a following amplifier via the coupling capacitor C2, only the *variations* in voltage are transferred, so that the final output signal is AC, as shown in (D).

Demodulating an FM signal is quite the opposite. Now the carrier frequency varies in accordance with the modulation, not the amplitude. To recover the audio signal you have to turn the frequency variations into voltage variations.

The simplest way to do this is with a slope detector, as shown in Figure D–10. The resonant circuit L1-C1, instead of being peaked at the center frequency of the IF signal, is tuned so that the center frequency is at some point on the slope of the resonant response curve—hence the name "slope detector."

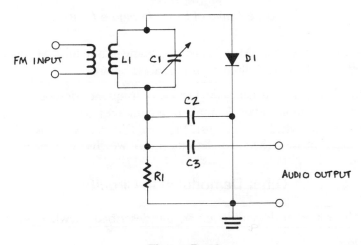

Figure D–10
Slope Detector

Figure D–11 shows the response curve of the L1-C1 resonant circuit. The 10.7-megahertz IF is at point A, and the slope is wide enough to permit a frequency deviation of 75 kilohertz from point A to B, and from A to C.

Obviously, an increase in the frequency of the FM signal will bring it closer to the peak of the response curve, so the amplitude of the signal applied to D1 will be greater. Conversely, a decrease will take it further away (down the slope), so the amplitude of the signal applied to the

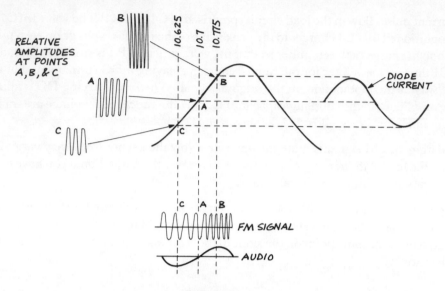

Figure D–11

Response Curve of L1-C1 in Figure D–10

diode will be less. In this way the FM signal can be changed into an AM signal, which merely has to be detected by a diode and filtered, as described before.

The slope detector clearly demonstrates the process of frequency demodulation, although it is not used in FM demodulation today. Its main disadvantages are that it is not linear enough, and it is sensitive to amplitude variations. Modern FM receivers use the ratio detector, quadrature detector, or phase-locked loop detector (q.v.), which overcome these disadvantages.

Other Demodulation Circuits

See the following list for other demodulation circuits described elsewhere in the encyclopedia:

Active detector
AM detector and audio amplifier
Converter
Crystal detector
Crystal set
De-emphasis
Delayed AVC
Discriminator
Dual-diode mixer
FET regenerative detector
Gated-beam detector
Grid-leak detector

Hot-carrier diode mixer
Infinite impedance detector
IC quadrature FM detector
Mixer
Pentagrid converter (see *Converter*)
Phase-locked loop detector
Plate detector
Quadrature FM detector
Ratio detector
Regenerative IGFET detector
Regenerative pentode detector
Solid-state detector and AVC
Squelch
Superheterodyne receiver
Transistor-amplified crystal detector
Transistor AM receiver
Transistor converter (see *Converter*)
Video detector

Demodulator Probe
(Figure D–12)

This is an oscilloscope probe for displaying the modulating signal on a high-frequency carrier. Parts values are indicated in the figure. See *Test and Indicating Circuits*.

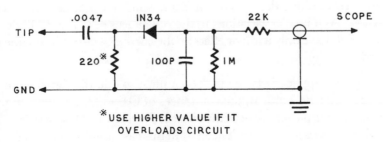

Figure D–12
Demodulator Probe

Differential Amplifier
(Figure D–13)

A differential amplifier consists of two identical amplifiers. If identical signals are applied to the input of each amplifier, identical amplified signals will appear at their outputs. A voltmeter connected between the two outputs will read zero. Such signals are called *common-mode* signals.

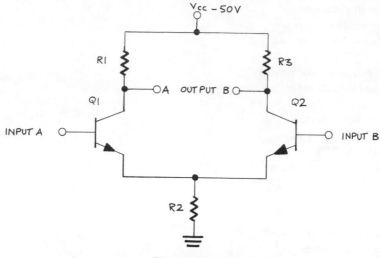

Figure D–13
Differential Amplifier

However, no two amplifiers can ever be *completely* identical, so there will always be a small percentage of the common-mode signal at the outputs. The size of this percentage is the measure of the amplifier's *common-mode rejection.*

The resistance of R2 is made quite large. On the other hand, the resistances offered by R1, Q1, R3 and Q2 are quite low, so that most of the supply voltage is dropped across R2. This gives it a dominant role in establishing the level of current flowing in the circuit, so it is regarded as a *constant-current source.* As a result, variations in the resistances of Q1 and Q2 due to undesired common-mode signals have only a minor effect on the current, which is the same as saying they are not amplified very much.

When an input signal is applied between the two inputs, so that the signal on input A is of opposite phase to the signal on input B, the resistances of Q1 and Q2 change in opposite directions. Now one transistor passes more of the current flowing in R2, so less is available for the other, which is all right because it doesn't need it. In this case amplification of the desired signal is enhanced.

From this you can see that R2 is a very important component, since it cuts down on the amplification of unwanted signals, but does not interfere with the amplification of legitimate ones. The circuit amplifies the *difference* between the voltages on the two inputs, while rejecting a voltage that is common to both.

Parts List

Q1, Q2	Matched pair of general-purpose NPN amplifier transistors
R1, R3	1.8 kilohms
R2	4.5 kilohms

Differentiator
(Figure D-14)

This circuit is used to convert retangular or other shaped pulses into spikes, such as trigger signals to initiate a time base, or sweep. Differentiators are used in TV receivers to separate and shape the horizontal sync pulses. In oscilloscopes they are used to change the trigger-multivibrator rectangular output into a series of negative and positive trigger spikes; since only one polarity is required, a diode is generally used to short out the unwanted spikes. The illustration shows a differentiator in a TV receiver.

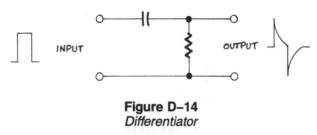

Figure D-14
Differentiator

Parts List

C1   0.001 microfarad
R1   100 kilohms

Diode Clipper

See *Clipper*.

Diode Detector

See *Crystal Detector*.

Dip Meter
(Figure D-15)

This circuit is used for measuring the resonant frequency of tuned circuits. The frequency of oscillation of Q1 is determined by the L1-C2 tank circuit. The oscillator signal is then rectified by D1, and the resulting DC is applied across M1. R4 is adjusted to give a full-scale reading.

To use the dip meter, L1 is positioned close to a resonant circuit whose resonant frequency is to be measured. C2 is then tuned until the meter reading dips. This results from RF energy being absorbed by the resonant circuit, which occurs because the two circuits are at the same frequency.

The frequency is indicated by the reading of C2's dial.

Several plug-in coils are provided in commercially available dip meters, each covering a specific frequency band. See *Test and Indicating Circuits*.

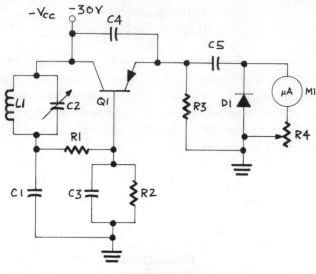

Figure D–15
Dip Meter

Parts List

C1	0.02 microfarad
C2	140 picofarads, variable
C3	0.02 microfarad
C4	4–7 picofarads
C5	27 picofarads
L1	For details on the construction of a set of coils, see *Absorption Frequency Meter*
M1	0–50 microamperes, DC
Q1	Any available RF/IF amplifier/oscillator PNP transistor
R1	39 kilohms
R2	3.9 kilohms
R3	270 ohms
R4	5 kilohms, variable

Direct-Coupled Amplifier (Figure D–16)

One of the advantages of solid-state circuits is the ease with which direct coupling can be made between stages. This removes the limitations on frequency response that result from the use of capacitors and transformers, making modern audio amplifiers more truly "hi-fi." This is the circuit of a 10-watt hi-fi audio amplifier, used in conjunction with a standard hi-fi preamplifier.

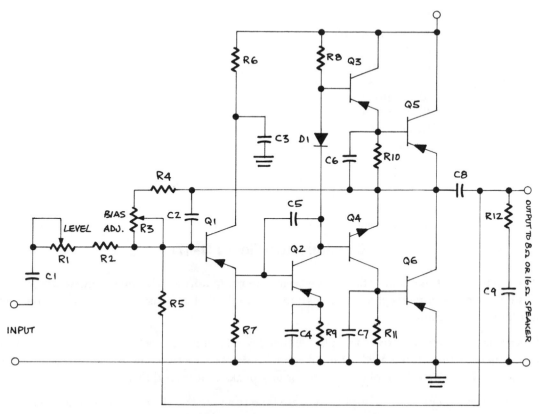

Figure D–16
Direct-Coupled Amplifier

Parts List

C1	20 microfarads, 20 volts
C2	220 picofarads
C3	25 microfarads, 50 volts
C4	100 microfarads
C5	0.001 microfarad
C6	0.005 microfarad
C7	0.005 microfarad
C8	1000 microfarads, 50 volts
C9	0.2 microfarad
D1	1N91*
Q1	2N320*
Q2, Q3	2N524*
Q4	2N167*

Q5, Q6	2N173*
R1	5 kilohms, variable
R2	1 kilohm
R3	100 kilohms
R4	150 kilohms
R5	20 kilohms
R6	39 kilohms
R7	1.5 kilohms
R8	8.2 kilohms
R9	470 ohms
R10, R11	1 kilohm
R12	22 ohms

*or equivalent

Discriminator (Figure D–17)

The full name of this circuit is the Foster-Seeley discriminator. Unlike the slope detector (q.v.), this FM detector converts phase changes into amplitude changes, instead of frequency changes into amplitude changes.

Since this detector is sensitive to amplitude changes, a limiter must be used ahead of it. V1 performs this function. It is biased to limit the amplitude of the IF signal to, say, 3 volts positive or negative by cutting off the plate current for any peaks that exceed this value.

The resonant circuits L2-C5 and L3-C7 are both tuned to the IF resting frequency, which is 10.7 megahertz in FM broadcasting and 4.5 megahertz in TV sound. This signal is inductively

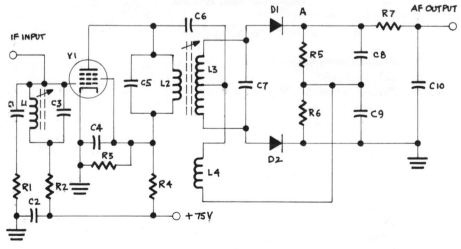

Figure D–17
Discriminator

coupled from L2 to L3. The IF signal voltage at the output of the limiter stage is also applied to L3 via C6.

The signal voltage coupled via C6 is also connected to L4. From L4 the connection is via C9 to ground, and then back to the limiter plate via R3 and L2. L2 and L4 are connected parallel as far as the RF signal voltage is concerned. Therefore the voltage across L4 is the same as that across L2.

The voltage induced in L3 divides between the center tap and either outside terminal to produce equal and opposite voltages. Since L3 is common to both circuits, the voltage across the upper half of L3 and that across L4 operate D1, while the voltage across the lower half of the coil and that across L4 operate D2. These signal voltages are rectified by the diodes and produce voltage drops across R5 and R6.

At resonance, a series resonant circuit acts as a resistance, and the voltage and current in a purely resistive circuit always have the same phase. So the voltage and current in L3 are in phase with each other. However, when the incoming FM signal deviates below the resonant frequency of the discriminator, the series resonant circuit L3–C7 becomes capacitive, and the induced current and voltage are no longer in phase. The current leads the voltage a certain amount, depending on the amount of frequency deviation. This causes the vector sums of the voltages across the two halves of L3 and of that across L4 to change. They are no longer equal, and they produce unequal voltages across R5 and R6. The *difference* between them, in this case negative, appears at point A.

The opposite happens when the incoming signal is above the resonant frequency. The circuit impedance is now largely inductive, so a phase shift in the opposite direction results. The difference between the voltages across R5 and R6 results in a positive voltage at point A.

To put it shortly:

(1) When the incoming signal is at the resting frequency, the voltage at A is zero.

(2) When the incoming signal is below the resting frequency, the voltage at A is negative.

(3) When the incoming signal is above the resting frequency, the voltage at A is positive.

The amount of the voltage at A corresponds to how much the frequency deviates; the rate at which it changes is the same as the rate at which the deviation changes. The voltage at A is, in fact, the audio signal recovered from the carrier. See *Demodulation Circuits.*

Parts List

C1, C3, L1	Limiter input resonant circuit (10.7 megahertz)
C2	5000 picofarads
C4	2000 picofarads

C5, C6, C7, L2, L3, L4	Limiter/discriminator resonant circuit
C8, C9	150 picofarads
C10	470 picofarads
D1, D2	1N64
R1	1 megohm
R2	22 kilohms
R3	47 kilohms
R4	10 kilohms
R5, R6, R7	100 kilohms
V1	12AU6

Doubler

See *Voltage Doubler*.

Double-Tuned Transformers
(Figure D-18)

Both the primaries and secondaries of double-tuned transformers are tuned by built-in variable capacitors (A), or variable inductance, using adjustable ferrite cores (B). These are standard for vacuum-tube circuits. However, IF-amplifier capacitors are not required in TV, as the distributed capacitance of the circuit is sufficient at the higher frequencies involved.

In transistor circuits, an interstage IF transformer not only has to select the IF signal with its tuned primary winding, but also has to act as a stepdown transformer to match the high impedance of the tuned circuit to the low or moderate input impedance of the following stage. As a general rule, only the transformer's primary winding is tuned. See *Amplifier Circuits*.

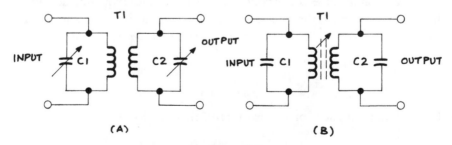

Figure D–18
Double-Tuned Transformers

DTL NAND Gate
(Figure D-19)

DTL stands for "diode-transistor logic." This circuit is really a combination of two circuits. D1, D2, and R1 are a simple diode AND gate, and Q1 and its associated components are an inverter to turn the AND into NAND. If the inputs of D1 and D2 are grounded, or at zero potential, they will be forward biased, because their cathodes will be less positive than their anodes. Since they are conducting, the point marked X will be at zero potential as well. Actually, the point will be at about 0.6 volt if the inputs are silicon diodes, because this is their forward conduction voltage, but this is generally called logic 0. So, with logic 0 on both gates (or either, for that matter), the output at X is also a logic 0.

If a logic 1 is applied to both gates, so that the input voltage is higher than the voltage at X, the diodes will be unable to conduct, so the voltage at X will rise to that of the supply voltage. This voltage will appear on the base of Q1, causing it to switch to the conducting state. Q1 goes from nonconducting to conducting, and its output goes from high to low. In other words, with both inputs high, the output is low, which is the NAND function.

As this gate is obtainable in IC form no parts list is given. See *Logic Circuits*.

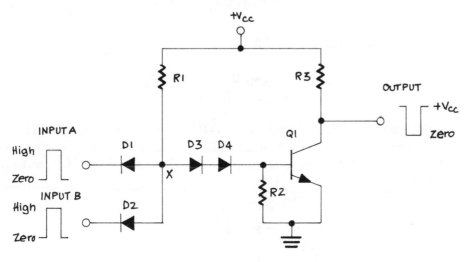

Figure D-19
DTL NAND Gate

Dual-Diode Mixer
(Figure D-20)

In this circuit the incoming RF signal is applied to the cathode of D1 and the anode of D2, while the local oscillator signal is applied to the anode of D1 and the cathode of D2 via T1. The local oscillator signal switches the diodes on and off alternately in accordance with the polarity of its cycles. The RF signal also affects the conduction of the diodes in a similar manner, but at a different rate. The resultant beat signals, together with the two originals, appear at the centertap of T1, from which the IF signal is selected by the following stage. See *Demodulation Circuits*.

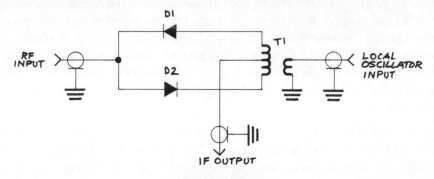

Figure D–20
Dual-Diode Mixer

Parts List

D1, D2	1N34A or equivalent
T1	RF transformer for required frequency

Dual-Voltage Power Supply
(Figure D–21)

In this circuit, T1 changes the line voltage to 6.3 volts RMS, and this is applied to D1 and D2, to be rectified. The resultant DC pulses appear across C1 and C2, which smooth out the ripple. S1 selects whether the output shall be the voltage across C1, or that across C1 and C2 combined. See *Power Supply Circuits*.

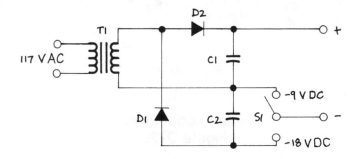

Figure D–21
Dual-Voltage Power Supply

Parts List

C1, C2   Any value between 100 and 1000 microfarads, 50 volts, depending on ripple and low
           current requirements
D1, D2   1N4000 series, or equivalent, 50 volts
S1   SPDT, any type
T1   Power transformer, 115 V/6.3 V (filament transformer would be suitable)

Dual Wavetrap
(Figure D-22)

In this circuit, L1 and C1 are a series-resonant wavetrap that provides a low-impedance path to ground for an unwanted signal to which they are tuned. At the same time, the parallel-resonant circuit L2 and C2 presents a high impedance to the same signal, if they are tuned to it.

While both wavetraps can be tuned to the same frequency for maximum efficiency against *one* interfering signal frequency, they can also be tuned to different frequencies to attenuate *two* interfering signals. Values of the inductors and capacitors will depend upon the frequency band of the receiver. For the broadcast band they would be as given below. See *Passive Filters*.

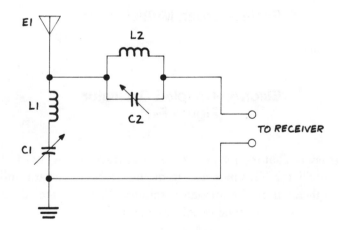

Figure D-22
Dual Wavetrap

Parts List

C1, C2   Single-section tuning capacitors, 30–350 picofarads
L1, L2   100 turns of No. 22 enameled copper wire wound on a 1-inch coil form (about 3 inches
           long)

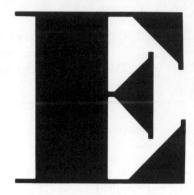

Eccles-Jordan Multivibrator

See *Bistable Multivibrator.*

Electron-Coupled Oscillator
(Figure E-1)

In this circuit a Hartley oscillator (q.v.) is used in an arrangement that isolates the output load circuit from the oscillator. This increases frequency stability and minimizes the adverse effect of variations in the load on the oscillator stability. The oscillator does not have to be a Hartley; the circuit can use any type of RF oscillator.

In this case, the oscillator consists of C1, L1, L2, R1, and C2, and the cathode, control grid, and screen grid of V1. The screen grid acts as the plate of a triode. The operation of the oscillator is the same as that described for the Hartley oscillator.

The screen grid modulates the electron flow from cathode to plate in V1, so that the oscillating signal in the Hartley oscillator is coupled to the plate by the electron flow variations. There is also some amplification. See *Oscillator and Signal-Generator Circuits.*

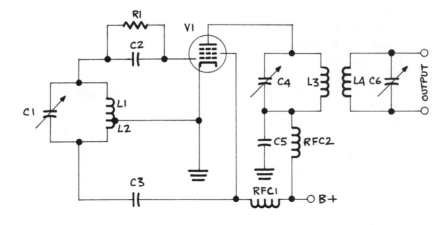

Figure E–1
Electron-Coupled Oscillator

Parts List

C1, C4, C6	Variable capacitors with capacitance that, together with the inductances L1-L2, L3 and L4, determine the resonant frequency of the circuit
C2	250 picofarads
C3	470 picofarads
C5	470 picofarads
R1	1 megohm
RFC1, RFC2	RF choke, 2.5 millihenries
V1	6AK6 or equivalent

Electronic Oven Control
(Figure E-2)

The circuit consists of two cascaded operational amplifers A1 and A2 driving a pair of transistors Q1 and Q2, which in turn control the current for the oven heater HR1. A diode D1 senses the oven temperature and controls A1, since it is in A1's feedback circuit. The potentiometer R3 is used to set the temperature to be maintained. See *Integrated Circuits*.

Parts List

A1, A2	Two halves of a μA739 operational amplifier pair, or equivalent
C1, C2, C3	0.01 microfarad

C4	10 picofarads
C5	5 nanofarads
D1	Temperature sensing diode
D2	Bias regulator diode (value depends on Q1 and Q2)
HR1	Oven heater element, 10 ohms
Q1, Q2	Power amplifier NPN transistors (I_C not less than 10 amperes)
R1	39 kilohms
R2	5.1 kilohms
R3	500 ohms, variable
R4	470 ohms
R5	18 kilohms
R6	620 kilohms
R7	2 kilohms
R8	4.7 kilohms
R9	Depends on transistors

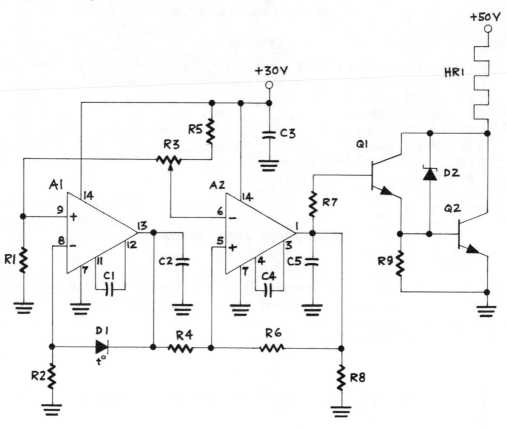

Figure E–2
Electronic Oven Control

Electron-Ray Tube Tuning Indicator
(Figure E-3)

This device, generally known as a ''magic eye,'' used to be popular in the better type of radio receiver, in the days before television when radio was king. It is still used in quite a few applications as a level indicator. In a radio receiver the AVC voltage is applied to the control grid of the magic eye. When the receiver is not tuned to a station, the AVC level will be very low, and the shadow angle of the magic eye will be about 100 degrees. When a station is tuned, the AVC voltage increases and the shadow angle decreases. Tuning for the narrowest shadow angle, therefore, ensures that the station is tuned for the maximum signal. See *Test and Indicating Circuits.*

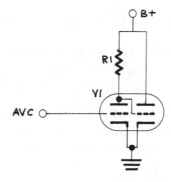

Figure E–3
Electron-Ray Tube Tuning Indicator

Parts List

R1   1 megohm
V1   6E5

Emphasis

See *Pre-emphasis* and *De-emphasis.*

Extension Speaker Feed System
(Figure E-4)

This circuit shows how to feed a local speaker and three remote speakers from the low-impedance output of an audio amplifier. If the load on the amplifier can be as low as 4 ohms, the impedance of all four speakers can be 8 ohms.

LS1 is the amplifier's local speaker. T1 is also connected across the amplifier output. It is a 70-volt speaker feedline transformer connected ''backwards.'' In other words, its secondary

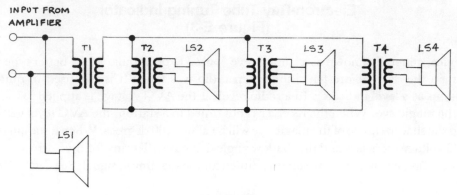

Figure E–4
External Speaker Feed System

is used as its primary, with the amplifier output connected between the common terminal and the 8-ohm tap. If the amplifier's average output is adjusted to deliver 10 watts to the 4-ohm load, its output voltage will be 6.3 volts RMS. T1 steps up this voltage to 70 volts.

Speakers LS2, LS3, and LS4 are connected to the 70-volt feedline by transformers similar to T1, but connected the right way round. The volume level of these speakers can be set independently by selecting the transformer secondary taps.

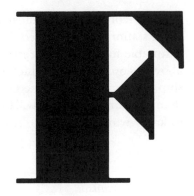

Ferrite-Core Antenna
(Figure F-1)

Ferrite is a ceramic-like compound of iron oxide and a metal, which exhibits weaker magnetism than iron, but is a much poorer conductor of electricity. This means it is not prone to losses from eddy currents, so it can be used to increase the effectiveness of transformers at higher frequencies than can be handled by ordinary iron-core transformers. Most RF and IF transformers are inductance tuned by a movable ferrite core.

A ferrite-core antenna, or loopstick, is a stepdown transformer. Two windings are wound on a form that surrounds the core. The larger primary winding is tuned by a variable capacitor to be resonant at the desired frequency. The smaller secondary winding is connected to the low-impedance input of the transistor converter or RF amplifier.

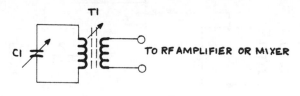

Figure F–1
Ferrite-Core Antenna

A ferrite-core antenna is most sensitive when either end of the core is pointed in the direction of the station being received. It is least sensitive when it is broadside to the station. This makes it suitable for low-cost direction finders, where a rotatable ferrite-core antenna is mounted above a compass rose, which may be oriented with respect to north, or to the heading of a ship. The antenna is rotated to obtain a null (weakest signal), since this is more pronounced than a maximum. There will be two nulls, 180 degrees apart, and it will be up to the navigator to know which is the correct one. See *Antenna Circuits*.

FET Drain Bias

See *Bias Circuits*.

FET Frequency Multiplier
(Figure F-2)

In this circuit, the cutoff bias and the amount of RF drive voltage determine the conduction angle of the FET (field-effect transistor), and hence the harmonic that will be generated by the multiplier. (The conduction angle is the segment of the electrical cycle during which drain current flows.) The input tank is tuned to the input frequency, and the output tank is tuned to the frequency of the desired harmonic. The harmonic numbers generated by various conduction angles are as follows:

CONDUCTION ANGLE DEGREES	HARMONIC NUMBER
90 – 120	2
80 – 120	3
70 – 90	4
69 – 72	5

See *Frequency Conversion Circuits*.

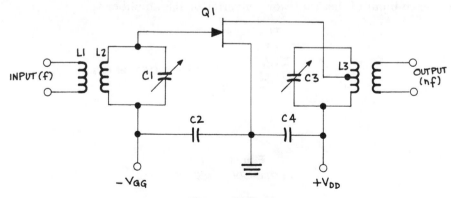

Figure F–2
FET Frequency Multiplier

Parts List

C1, C3	Depends on frequency
C2, C4	100 picofarads
L1, L2, L3	Depends on frequency
Q1	N-channel JFET (any general-purpose type)

FET Oscillator and Frequency Doubler
(Figure F-3)

In this circuit, Q1 is an oscillator generating a signal frequency of 14 megahertz. This signal is coupled through T1 to Q2 and Q3, which operate in push-push to double the frequency to 28 megahertz. R3 is for balancing the drain currents of Q2 and Q3. See *Frequency Conversion Circuits*.

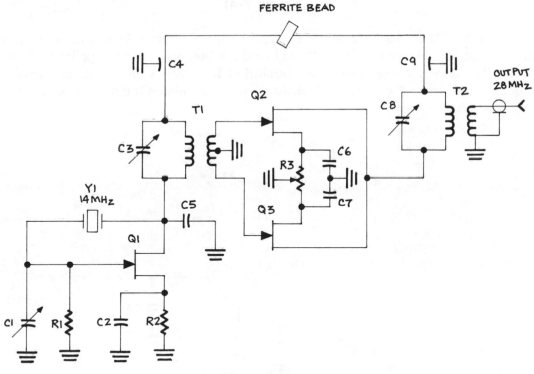

Figure F–3
FET Oscillator and Frequency Doubler

Parts List

C1	10–60 picofarads, variable
C2, C6,	0.01 microfarad

C7	
C3	3–12 picofarads, variable
C4, C9	1000 picofarads
C5	100 picofarads
C8	8–35 picofarads, variable
Q1, Q2, Q3	N-channel JFETs, MPF102/HEP802, or equivalent
R1	1 megohm
R2	1 kilohm
R3	2 kilohms, variable
T1	Interstage transformer for 14 megahertz
T2	Interstage transformer for 28 megahertz
Y1	14-megahertz crystal

FET Regenerative Detector
(Figure F-4)

In this circuit, Q1 is used in a modified Hartley regenerative circuit. Incoming radio signals are amplified and demodulated by Q1, and feedback from its source is supplied to the tap on L2. The amount of regeneration is controlled by R2, which varies the drain voltage. It should be adjusted so that the circuit is on the verge of oscillation. C1 is the tuning capacitor. See *Demodulation Circuits*.

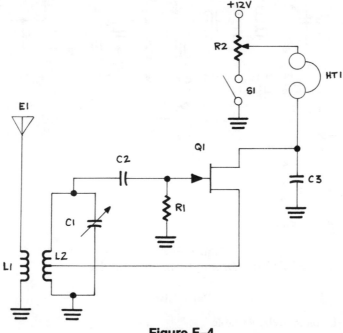

Figure F–4
FET Regenerative Detector

Parts List

C1	15 picofarads, variable
C2	100 picofarads
C3	0.001 microfarad
L1, L2	According to frequency band: for 28 megahertz, L1 should consist of 4 turns of No. 30 enameled copper wire, and L2 should consist of 8 turns of No. 24 enameled copper wire, both close-wound on a Millen 74001 1/2 inch turnable coil form. The tap should be on the second turn above ground.
Q1	N-channel JFET, HEP801, or equivalent
R1	22 kilohms
R2	50 kilohms, variable

Field Strength Meter
(Figure F-5)

This circuit consists of two sections. The first is the pickup unit, which has a rod antenna about 24 inches long (less for high radiated power), tuning circuits for three bands, a detector diode, and a plug. The second, which may be directly connected by plugging the first into its socket, consists mainly of a microammeter. The reason for the plug and socket is so that the units may be connected by an extension cord, which can be any length up to several hundred feet. This arrangement allows the pickup unit to be set up out in the field to pick up radiation from the antenna being tested, while the meter unit is near where the adjustments are to be made. See *Test and Indicating Instruments.*

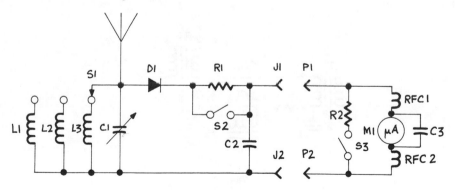

Figure F–5
Field Strength Meter

Parts List

C1	25 picofarads, midget variable
C2, C3	0.001 microfarad, mica
D1	1N34 or equivalent
J1, J2	AC sockets, two pin, preferably polarized*

L1	*28-MHz coil:* 7 turns of No. 22 enameled copper wire, 1/4 inch long on a 3/4-inch diameter form
	50-MHz coil: 6 turns of No. 22 enameled copper wire, 1/4 inch long on a 9/16-inch diameter form
	144-MHz coil: 3 turns of No. 18 enameled copper wire, 1/4 inch long, 3/8-inch diameter, self-supporting
M1	0–100 microammeter (less sensitive meters may be used with reduced unit sensitivity)
P1, P2	AC plugs, two pin, preferably polarized*
R1	1 kilohm
R2	220 ohms
RFC1, RFC2	2.5 millihenries (RF choke)
S1	Three-position wafer switch
S2, S3	SPST toggle switches

*One should be for chassis mounting.

Filament-Type Tube AF Amplifier
(Figure F-6)

This type of circuit is seen only in very old equipment dating from before the invention of indirectly-heated cathodes. The center tap on the filament transformer divided the winding into two halves, so that they would buck the AC hum. No parts are listed for this circuit. See *Amplifier Circuits*.

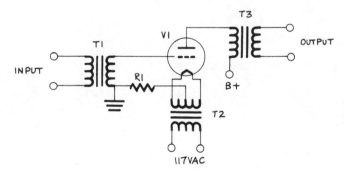

Figure F–6
Filament-Type Tube AF Amplifier

Fixed Bias

See *Bias Circuits*.

Flip-Flop

See *Bistable Multivibrator.*

Franklin Oscillator
(Figure F-7)

This circuit is a two-stage amplifier with positive feedback from the output of the second stage to the input of the first. It is very stable and is suitable for use as a frequency-determining stage, but not as a power oscillator. See *Oscillator and Signal-Generator Circuits.*

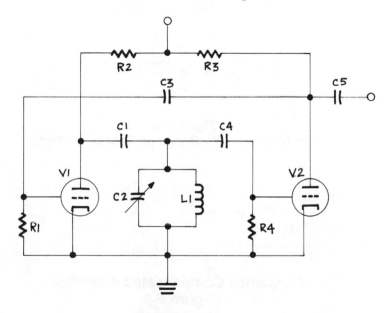

Figure F-7
Franklin Oscillator

Parts List

C1, C4	47 picofarads
C2	Tuning capacitor, value depends on frequency band
C3, C5	0.02 microfarad
L1	Tank coil, value depends on frequency band
R1	1 megohm
R2, R3	100 kilohms
R4	220 kilohms
V1, V2	6C4

Free-Running Pulse Repetition Rate Generator
(Figure F-8)

This is a blocking oscillator circuit (q.v.), from which short, high-energy pulses are obtained at a rate determined by the RC time constant of R1 and C1. The pulse width depends upon the inductance of T1. See *Oscillator and Signal-Generator Circuits*.

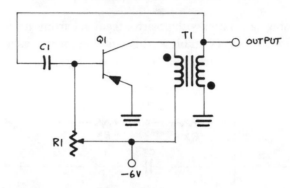

Figure F–8
Free-Running Pulse Repetition Rate Generator

Parts List

C1   10 microfarads
Q1   General-purpose PNP transistor
R1   500 kilohms, variable
T1   Blocking oscillator transformer

Frequency-Compensated Amplifier
(Figure F-9)

This circuit shows a method for increasing the high frequency gain of an amplifier such as a video amplifier (q.v.). The impedance of the peaking coil L1 increases with frequency, offsetting the effect of shunt capacitance in the tube. See *Amplifier Circuits*.

Parts List

C1, C3   0.1 microfarad
C2   1 microfarad
L1   300 microhenries (depends on frequency)
R1   1 megohm
R2   5 kilohms
R3   200 kilohms
V1   6C4

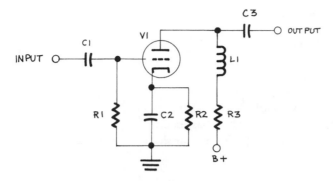

Figure F–9
Frequency-Compensated Amplifier

Frequency-Compensated Preamplifier
(Figure F-10)

In this preamplifier circuit the frequency response can be changed by the setting of S1, the frequency compensation selector switch. Controls for input level, treble equalization, bass boost, and level are also provided. See *Amplifier Circuits.*

Parts List

C1, C3	0.1 microfarad
C2	20 microfarads, 15 volts
C4	0.0082 microfarad
C5	0.0047 microfarad
C6	0.0068 microfarad
C7	200 microfarads
C8	0.027 microfarad
C9	100 microfarads
C10	20 microfarads, 15 volts
L1	0.7-henry reactor
Q1, Q2	General-purpose PNP transistors
Q3	
R1	500 kilohms, variable ("input level")
R2	330 kilohms
R3	56 kilohms
R4	100 kilohms
R5	18 kilohms
R6	25 kilohms, variable ("treble equalization")
R7	20 kilohms
R8	56 ohms
R9	47 ohms

R10	1.5 kilohms
R11	270 kilohms
R12	11 kilohms
R13	10 kilohms
R14	82 ohms
R15	2.7 kilohms
R16	5 kilohms, variable (''level'')
R17	50 kilohms, variable (''bass boost'')
S1	Two-pole, four-position, non-shorting rotary switch

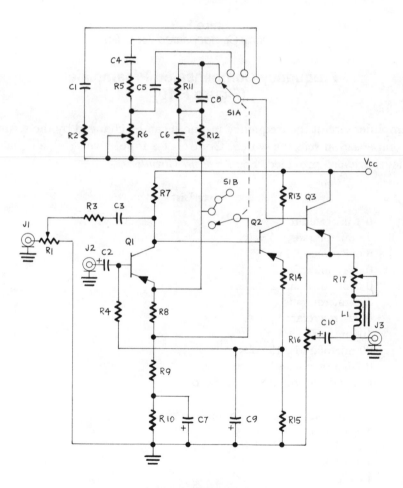

Figure F–10
Frequency-Compensated Preamplifier

FREQUENCY CONVERSION CIRCUITS

Frequency conversion can be where you multiply or divide the original frequency by whole numbers, or where you mix frequencies to obtain "beat" frequencies, which seldom bear a whole-number relationship to the original frequencies.

Frequency multiplication is often used in transmitters, because it becomes increasingly difficult to maintain oscillator stability as the frequency is increased. The usual practice in working at the higher frequencies is to operate the oscillator at a low frequency and follow it with one or more frequency multipliers as required to arrive at the desired output frequency. An example of this is shown in Figure F-11, which is the circuit of a narrow-band FM transmitter. V1 is a crystal oscillator, in which plug-in crystals ranging in frequency from 937.5 to 1250 kilohertz are used. V2 and V3 are two balanced modulators (q.v.) which receive an audio signal from T1 (Figure B-3). The output of the modulator stage is then frequency-multiplied by two quadrupler stages V4 and V5, and by a doubler V6, to give a total multiplication of 32 times. All three of these tubes act as Class-C amplifiers with their plate tanks tuned to either the second or fourth harmonic of their grid signals. The final frequency is in the 30 to 40 megahertz range. It will be exactly 32 times the frequency of the crystal used in the oscillator circuit.

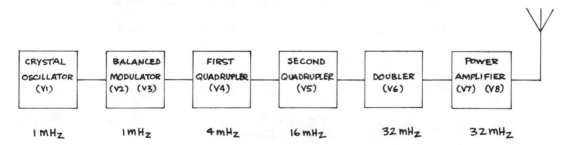

Figure F-11
Frequency Multiplication by Quadrupler and Doubler Stages

Dividing frequency is a common requirement of digital circuits. For instance, the crystal oscillator in a digital watch has a high frequency which is divided down to seconds, minutes, and hours by a series of cascaded dividers. These dividers are also termed counters, because they are often used for binary counting.

Figure F-12 shows how flip-flops, or bistable multivibrators (q.v.), are cascaded to form a simple frequency divider. This particular type of flip-flop is called a JK flip-flop. A JK flip-flop with both J and K inputs high (logical 1) changes the state of its output each time its clock input goes from 1 to 0. A series of pulses is applied therefore to the clock input of the first flip-flop. When the first pulse goes from 1 to 0, the Q1 output (initially 0) goes to 1. When the second pulse goes from 1 to 0, Q1 goes back to 0, and this causes Q2 to go to 1. When the third pulse goes from 1 to 0, Q1 goes to 1 again. When the fourth pulse goes to 0, Q1 goes to 0, causing Q2 to go to 0, which causes Q4 to go to 1. When the fifth pulse goes to 0, Q1 goes to 1. When the sixth pulse goes to 0, Q1 goes to 0, and Q2 goes to 1. When the seventh pulse goes to 0, Q1 goes to 1. When the eighth pulse goes to 0, Q1 goes to 0, Q2 goes to 0, Q4 goes to 0, and Q8 goes to 1. This means that eight pulses at the input of the first flip-flop have started one pulse at the output of the fourth flip-flop (another eight will be required to end it), so the original rate, or frequency, has been divided sixteen times.

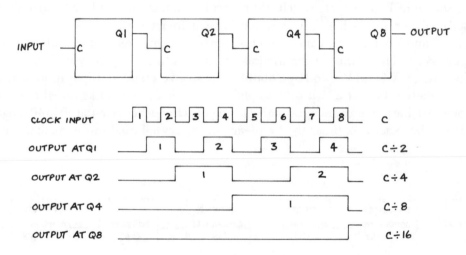

Figure F–12
Frequency Divider Using Cascaded Flip-Flops

Frequency division can therefore be accomplished to any integral quotient by cascading the correct number of flip-flops. (The same arrangement is used in binary counters, except that each Q output drives a readout.)

Frequency conversion by beating one signal with another is familiar to most in the form of the heterodyne process. As employed in modern receivers, the frequency of the incoming signal is changed to a new radio signal at the intermediate frequency (IF) for further amplification and detection. In this process, the output of a tunable oscillator, generally referred to as the local oscillator, is combined with the incoming signal in a mixer or converter stage, sometimes called the first detector, to produce a beat frequency equal to the IF.

As a numerical example, assume that the IF is 455 kilohertz and the incoming signal frequency is 7 megahertz. With the local oscillator frequency set at 7.455 megahertz, a beat frequency of 7,455 − 7,000 = 455 kilohertz is obtained. Another beat frequency at 7,455 + 7,000 = 14,455 kilohertz is also generated, but as the IF stage is not tuned to this frequency it is of no importance.

Each local oscillator frequency will cause IF response at two incoming signal frequencies, one higher and one lower than the oscillator frequency. With the oscillator set as above to 7.455 megahertz, the receiver can respond to a signal with a frequency of 7.910 megahertz, which likewise gives a 455-kilohertz beat. The resultant undesired signal is called an image. The degree to which such images are rejected depends upon the selectivity of the receiver's tuning circuits ahead of the mixer or converter stage, and the IF itself, since the higher the IF the greater the frequency separation between the desired signal and the image, placing the latter further away from the resonance peak of the signal-frequency input circuits.

The heterodyne process may be used for obtaining any desired frequency from another by generating a second signal to beat with it.

Other Frequency Conversion Circuits

See the following list for other frequency conversion circuits described elsewhere in the encyclopedia:

Converter
FET frequency multiplier
FET oscillator and frequency doubler
Frequency converter
Frequency multiplier
Frequency-multiplier trap
Frequency-multiplier trap with dual pi-networks
Hot carrier diode mixer
Low-impedance frequency doubler
Mixer
Oscillator and frequency doubler
Pentagrid converter
Push-pull frequency multiplier (transistor)
Push-pull frequency multiplier (tube)

Frequency Converter
(Figure F-13)

This circuit enables an AM radio to receive transmissions in the 27-megahertz citizen's band. A local oscillator with a 28-megahertz crystal heterodynes with the incoming 27-megahertz band signal to give a beat frequency of approximately 1000 kilohertz. This is applied to the antenna terminals of an AM radio, which can then be tuned for reception of transmissions in the band. The circuit as shown here was designed for use with a car radio, but could easily be modified for use with other equipment. The power supply could be a single flashlight cell. See *Frequency Conversion Circuits*.

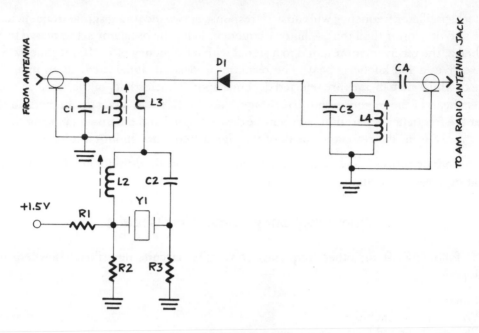

Figure F–13
Frequency Converter

Parts List

C1, C2	56 picofarads
C3	150 picofarads
C4	1000 picofarads
D1	Tunnel diode
L1, L2	0.57 microhenry
L3	Pickup coil, 2 turns
L4	170 microhenries
R1	750 ohms
R2, R3	100 ohms
Y1	28-megahertz crystal

Frequency-Modulated IF Amplifier
with Integrated Circuit
(Figure F-14)

This IF amplifier circuit uses a single IC between two IF transformers. The IF can be 4.5 or 10.7 megahertz (television sound or FM radio), using the appropriate transformers. For TV the gain will be 70 decibels, for radio 50 decibels. See *Amplifier Circuits; Integrated Circuits*.

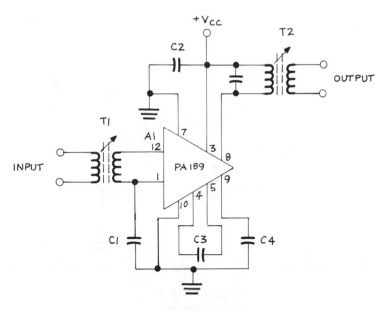

Figure F–14
FM IF Amplifier with IC

Parts List

A1	Integrated circuit PA189
C1, C3, C5	0.005 microfarad
C2	0.05 microfarad
T1, T2	IF transformers

Frequency-Modulated Ratio Detector

See *Ratio Detector*.

Frequency-Modulated Modulator
(Figure F-15)

Frequency modulation of a crystal oscillator is obtained in this circuit by connecting a varactor in series with the crystal. When an audio signal is applied to the varactor its capacitance varies in response to variations in the signal amplitude, so that the oscillator frequency varies accordingly. See *Modulation Circuits*.

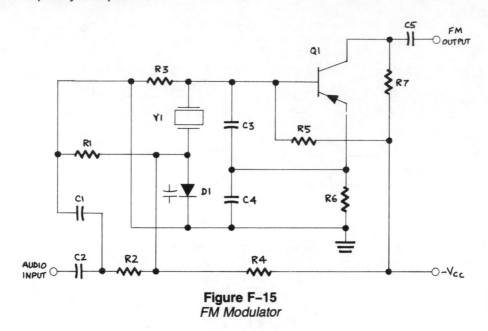

Figure F–15
FM Modulator

Parts List

C1, C2	1.0 microfarad
C3	700 picofarads
C4	0.0021 microfarad
C5	220 picofarads
D1	Varactor diode
Q1	GE82 or equivalent
R1, R2	1 kilohm
R3, R5	10 kilohms
R4	22 kilohms
R6	1 kilohm
R7	5.6 kilohms
Y1	3.5-megahertz crystal

Frequency Multiplier
(Figure F-16)

This circuit employs a single parallel-resonant output circuit that is tuned to a harmonic of the RF input. The input pi-network is tuned to the input frequency so that it presents the correct impedance to the signal source. The output tank coil is tapped down to allow interfacing with the low impedance load of the transmission line.

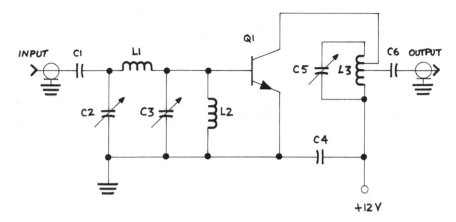

Figure F–16
Frequency Multiplier

The transistor is biased for nonlinear operation. In fact, it operates as a Class-C amplifier. This enables it to generate harmonics of the input frequency.

When no input signal is present there is no forward bias on Q1, since the base and emitter are shortened together through L2, which has practically no resistance for DC. When an input signal is present the choke has a relatively high reactance (3563 ohms at 21 megahertz). When the input signal makes Q1's base positive, the transistor is forward biased, and collector current flows through the tank coil. On the other hand, when the input signal makes Q1's base negative, the transistor is reverse biased, and no collector current flows. The output current flows in a series of powerful pulses that shock-excite the tank circuit into oscillation. As this is an LC circuit, its output waveform is a sine wave. The tank circuit is tuned to the desired harmonic (42 megahertz with an input of 21 megahertz, if the second harmonic is selected). See *Frequency Conversion Circuits*.

Parts List

C1, C4	1000 picofarads
C2	170–780 picofarads, variable
C3	100–580 picofarads, variable
C5	100 picofarads, variable
C6	100 picofarads, mica
L1	0.21 microhenry
L2	27 microhenries (RF choke)
L3	14 turns of No. 24 enameled copper wire, wound on a 1-inch coil form to make a coil 3/8 inch long; taps to match impedance (total inductance is approximately 6 microhenries) of transistor and transmission line
Q1	2N743

Frequency Multiplier Trap
(Figure F-17)

The use of a series-resonant wavetrap at the input side of a transistor frequency multiplier can improve the multiplier's efficiency. The input tank is tuned to the frequency of the input signal, but the wavetrap connected between the base and emitter of the transistor is tuned to the output frequency. This results in minimizing the effect of the base-to-collector capacitance. No parts values are given, as they depend upon signal frequencies to be handled. L1 and C3 are the series trap tuned to the output frequency. See *Frequency Conversion Circuits*.

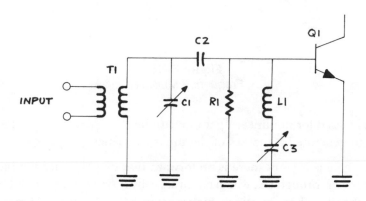

Figure F–17
Frequency Multiplier Trap

Frequency Multiplier with Dual Pi-Networks
(Figure F-18)

This circuit has both input and output pi-networks. The input pi-network, consisting of L1, C2, and C3, is tuned to the input signal frequency. The output pi-network, consisting of L4, C6, and C7, is tuned to the desired harmonic frequency. When used as a frequency doubler, a wavetrap (L3, C4, and C5) is required to remove the input signal frequency from the output. If the circuit is used as a frequency tripler, two wavetraps are necessary, one for the input signal and one for the second harmonic. This circit is designed for an input frequency of 21 megahertz, and an output of 42 megahertz, and to work into a 50-ohm load. See *Frequency Conversion Circuits*.

Parts List

C1, C8, C9	0.001 microfarad
C2	170–780 picofarads, variable
C3	100–560 picofarads, variable

C4	22 picofarads
C5	2.7–30 picofarads, variable
C6, C7	6.0–80 picofarads, variable
L1	0.21 microhenry
L2	27 microhenries (RF choke)
L3	287 microhenries
L4	0.086 microhenry
L5	10 microhenries (RF choke)
Q1	2N743 (shielded)

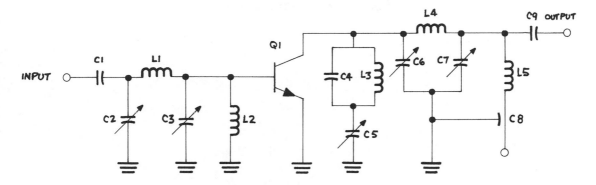

Figure F–18
Frequency Multiplier with Dual Pi-Networks

Full-Wave Rectifier
(Figure F-19)

There are two principal types of full-wave rectifier: One is the bridge rectifier (q.v.), and the other the full-wave centertap shown in Figure F-19.

Line voltage is applied to the primary of T1, and is converted to AC at some other voltage according to the turns ratio of the transformer. During the AC half-cycle when the upper end of T1's secondary is positive, electron current flows from the centertap, through the load, and via D1 to the upper end of the secondary. During the other half-cycle, when the lower end of the secondary is positive, electron current flows from the centertap, through the load, and via D2 to the lower end of the secondary. Polarity of the voltage across the load will be as shown. This voltage consists of a series of pulses, one for each positive half-cycle, and one for each negative half-cycle of input voltage. The ripple frequency will be twice that of the line voltage. No parts values are given for this circuit, as they will depend upon the design of the complete power supply, of which this is merely the rectifier section. See *Power Supply Circuits*.

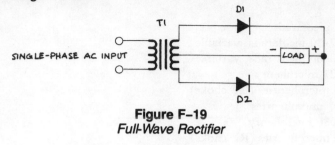

Figure F–19
Full-Wave Rectifier

Full-Wave Voltage Doubler

See *Bridge Voltage Doubler*.

Gain Controls
(Figures G-1 and G-2)

The circuit in Figure G-1 shows how the gain of a remote cutoff pentode is controlled by a variable resistor in the cathode circuit. This resistor varies the cathode bias by altering the combined resistance of R1 and R2, and so varies the gain of the tube. If the tube is a sharp cutoff pentode, its gain is varied by changing the screen-grid voltage by means of R2, as shown in Figure G-2. See *Amplifier Circuits*.

Parts List (Figure G-1)

C1, C4	0.01 microfarad
C2	12 microfarads
C3	0.1 microfarad
R1	500 ohms, 1 watt
R2	500 ohms, variable, 1 watt
R3	270 kilohms
R4	680 kilohms
V1	6BA6

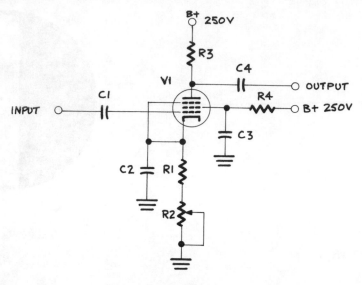

Figure G–1
Gain Control for Remote Cutoff Pentode

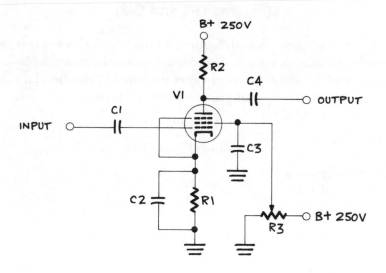

Figure G–2
Gain Control for Sharp Cutoff Pentode

Parts List (Figure G-2)

C1, C4	0.01 microfarad
C2	12 microfarads
C3	0.1 microfarad

R1   1 kilohm, 1/2 watt
R2   220 kilohms
R3   500 kilohms, variable
V1   6AU6

Gated-Beam Detector
(Figure G-3)

This FM demodulation circuit uses a special tube in which, in addition to the cathode and plate, there are three electrodes called the "signal" or "limiter" grid, the "accelerator" grid, and the "quadrature" grid. There are also various shields to confine the electron stream and form it into a beam.

Electrons from the cathode must flow through the openings in the accelerator grid, and past the signal and quadrature grids to reach the plate. The positive potential on the accelerator electrode attracts electrons away from the cathode and concentrates them into a narrow vertical beam.

The bias on the signal and quadrature grids is adjusted by R2 so that a very small signal voltage will drive either grid between cutoff and plate current saturation. Thus, if the signal grid is more than a few volts negative, the beam cannot pass through the accelerator. On the other hand, if the signal grid bias is somewhere between plate current cutoff and plate current saturation, the electrons will pass on toward the quadrature grid.

The quadrature grid acts in the same way as the signal grid, so both grids can be considered as gates to control the flow of electrons between the cathode and the plate. Since only a very small signal variation, usually about 1.25 volts, is needed to drive the tube between cutoff and saturation, amplitude variations on the positive and negative peaks of the incoming signal will be removed, so the tube also acts as a limiter.

The IF transformer T1 is tuned to the center frequency of the IF signal, and it couples the FM signal to the signal grid of V1. The sharp cutoff characteristic of the signal grid causes the plate current of the tube to vary from cutoff to full current and back once each cycle. The plate current is therefore in the form of pulses.

These current pulses passing through the tube shock-excite L1-C3. This circuit is also tuned to the IF center frequency. Because of the method of exciting this circuit, the voltage pulse across C3 (and on the quadrature grid as well) is 90 degrees out of phase with the current passing through the tube.

When the incoming signal is at the resting frequency (no modulation) there is a 50 percent overlap during which time both grids can pass the electrons through to the plate. If the frequency of the signal deviates above the center frequency, L1-C3 become reactive, and the overlap decreases in proportion to the deviation. On the other hand, when the FM signal deviates below the center frequency, the overlap increases above 50 percent in proportion to the deviation.

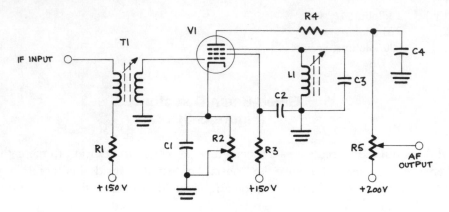

Figure G–3
Gated-Beam Detector

The plate current pulses across R4 charge C4 to the average of the pulse peaks, so the average voltage varies according to the frequency deviations in the FM signal. The longer pulses charge the capacitor above the resting frequency voltage, and the shorter pulses charge it less than the resting frequency voltage. This capacitor also grounds the IF frequency, and the RC time constant of R4 and C4 provides the proper de-emphasis (q.v.). R5 is the volume control. See *Demodulation Circuits.*

Parts List

C1	0.01 microfarad
C2	0.005 microfarad
C3	15 picofarads
C4	0.001 microfarad
L1	83 microhenries
R1	10 kilohms
R2	500 ohms (''buzz control'')
R3	56 kilohms
R4	470 ohms
R5	330 kilohms (''volume control'')
T1	4.5 megahertz IF transformer
V1	6BN6

Grid-Dip Meter

See *Dip Meter.*

Grid-Leak Bias

See *Grid-Leak Detector.*

Grid-Leak Detector
(Figure G-4)

In this circuit, V1 is a combination diode rectifier and audio-frequency amplifier. Its grid and cathode correspond to the anode and cathode of a diode detector, and the rectifying action is the same as previously described under *Crystal Detector* (q.v.).

The DC voltage from the rectified-current flow through the grid leak R1 biases the grid negatively with respect to the cathode; the audio-frequency variations in voltage across R1 are amplified by V1 as in any other amplifier. R2 is the plate load resistance, C3 bypasses the RF to ground, and the RF choke L3 prevents it from getting into the output circuit. See *Demodulation Circuits.*

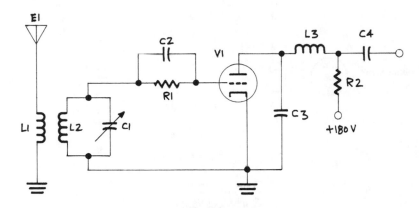

Figure G–4
Grid-Leak Detector

Parts List

C1   15 picofarads, variable
C2   100-250 picofarads (according to frequency band)
C3   0.001 microfarad
C4   0.1 microfarad
L1   28 MHz: 4 turns of No. 30 enameled copper wire*
L2   28 MHz: 8 turns of No. 24 enameled copper wire*
L3   2.5 millihenries
R1   1–2 megohms (according to frequency band)
R2   50 kilohms
V1   6C4

*Both close-wound on a Millen 74001 coil form (1 inch in diameter)

Grid Modulation
(Figure G-5)

This circuit is a typical arrangement for grid-bias modulation. The C– grid supply current flows through the secondary of T1. The primary of T1 is connected in the plate circuit of the modulator tube, so the audio voltage from the modulator varies the grid bias on V2, which is operated as a Class-C amplifier. F-F are the filament supply terminals. See *Modulation Circuits*.

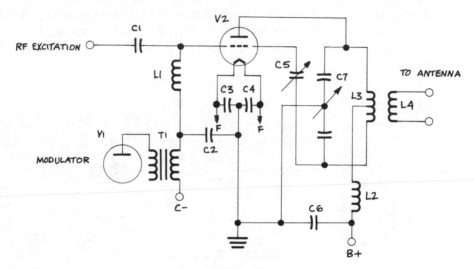

Figure G–5
Grid Modulation

Parts List

C1	47 picofarads, 1000 volts
C2	0.005 microfarad
C3, C4	0.01 microfarad
C5	Neutralizing capacitor (National STN or equivalent)
C6	0.01 microfarad
C7	100 picofarads per section, transmitting variable
L1	2.5 millihenries, 100 milliamperes (RF choke)
L2	Transmitting RF choke (Millen 34140 or equivalent)
L3	10 turns of No. 22 enameled copper wire, 1 inch diameter, 1–1/8 inches long, centertapped (14 and 21 megahertz)
L4	3-turn antenna link coil
T1	Audio output transformer
V1	Modulator output, 6L6 or equivalent
V2	5514

Grounded-Anode Amplifier

See *Grounded-Plate Amplifier*.

Grounded-Base Amplifier

See *Common-Base Transistor Amplifier*.

Grounded-Cathode Amplifier
(Figure G-6)

This circuit is the most widely used vacuum-tube circuit. The signal input is to the grid and the output is from the plate. The main advantages of this connection are maximum voltage gain, high input impedance at the grid, and substantial voltage and current output from the plate.

Since the audio amplifier with a triode (q.v.) has already been described, the circuit shown here illustrates an audio amplifier using a pentode. See *Amplifier Circuits*.

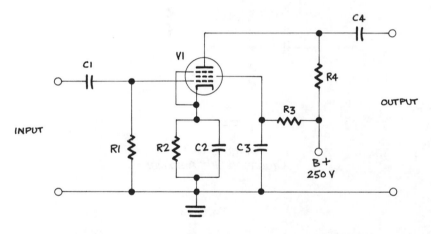

Figure G-6
Grounded-Cathode Amplifier

Parts List

C1, C4	0.05 microfarad
C2	20 microfarads
C3	0.022 microfarad
R1	270 kilohms
R2	1 kilohm
R3	680 kilohms
R4	270 kilohms
V1	6AU6

Grounded-Grid Amplifier
(Figure G-7)

In this circuit, the input signal is applied between the cathode and ground, and the output is taken from the plate. The grid is grounded, either directly as shown, or via a capacitor that gives a signal ground while maintaining a desired bias on the grid.

This circuit is frequently used in high-frequency amplifiers, where a high grid impedance would present a mismatch for low impedance signals. In the grounded-grid connection, the control grid also serves as an electrostatic shield between the cathode and the plate, thereby reducing the capacitance between them. One of the features of the grounded-grid circuit is that, unlike the grounded-cathode circuit, there is no reversal of polarity between input and output.

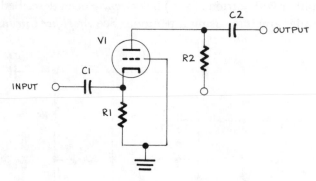

Figure G–7
Grounded-Grid Amplifier

Parts List

C1, C2	10 picofarads
R1	75 ohms (or whatever impedance is to be matched)
R2	100 kilohms
V1	6C4

Grounded-Plate Amplifier

See *Cathode Follower.*

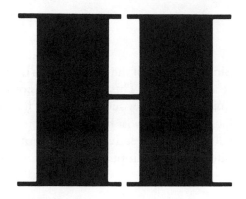

Half-Wave Doubler
(Figure H-1)

In this circuit, often called the "conventional" voltage doubler to distinguish it from the cascade voltage doubler (q.v.), capacitors C1 and C2 are each charged, during alternate half-cycles, to the peak value of the alternating input voltage. The capacitors are discharged in series into the load, thus producing across it an output of approximately twice the AC peak voltage. Component values will depend upon the load, type of filter used, and input and output voltages. See *Power Supply Circuits*.

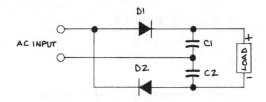

Figure H–1
Half-Wave Doubler

153

Half-Wave Rectifier
(Figure H-2)

Since only half of the input wave is used, the efficiency of this circuit is low and the regulation relatively poor. This circuit does not work very well with a transformer input, because the unidirectional secondary current flow tends to cause core saturation and poor regulation. Consequently most half-wave circuits operate either directly from the AC line, or at a high voltage with a relatively low current. Component values will depend upon the load, type of filter used, and input and output voltages. See *Power Supply Circuits.*

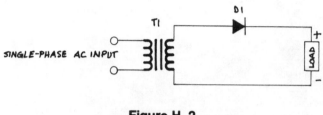

Figure H–2
Half-Wave Rectifier

Hartley Crystal-Controlled Oscillator
(Figure H-3)

In this circuit, the only difference between the crystal-controlled version and the standard oscillator (q.v.) is the addition of a crystal in the feedback circuit. This has the effect of blocking feedback except at the crystal frequency, making it impossible for the circuit to oscillate at any other frequency. See *Oscillator and Signal-Generator Circuits.*

Parts List

C1	33 picofarads
C2, C3	0.01 microfarad
L1	4 microhenries
Q1	GE20, or equivalent general-purpose NPN transistor
R1	22 kilohms
R2	4.7 kilohms
R3	47 ohms
Y1	14-megahertz crystal

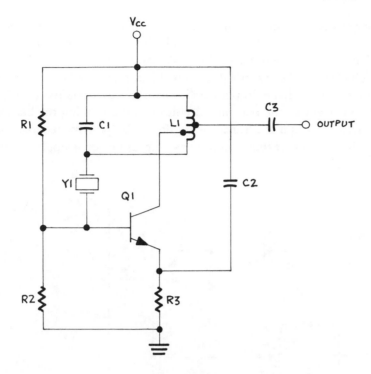

Figure H-3
Hartley Crystal-Controlled Oscillator

Hartley Oscillator
(Figure H-4)

This circuit is a standard Hartley oscillator. The resonant circuit consists of a tapped inductor L1–L2 and a variable capacitor C1. The tap splits the inductor into two sections. L1 is the base-emitter section; L2 is the emitter-collector section.

When power is applied to the circuit, electron flow begins to build up, starting from ground and proceeding via L2 to Q1's emitter. As it builds up it induces a voltage in L1 such that the side of C1 connected to C2 becomes positively charged. However, C2's reactance at the frequency of oscillation is very low, so this positive charge also appears on the base of Q1.

This increases the forward bias across the emitter-base junction, causing an increase in the collector current of Q1. This results in a still further increase in the electron flow in L2.

This burst of energy is more than enough to start the resonant circuit L1–L2 and C1 oscillating. Consequently, the end of L1 connected to C2 will alternate between positive and negative. You've seen what happens when it is positive. When it goes negative, the negative voltage is coupled through C2 on to the base of Q1, cutting the transistor off. C2 now starts to discharge through R1, setting up a voltage drop across it such that Q1 remains cut off. This situation lasts until the polarity across the resonant circuit reverses, making the base positive again. Each burst of energy through the transistor reinforces by positive feedback the oscillations in the resonant circuit, so that they continue as long as power is applied. See *Oscillator and Signal-Generator Circuits.*

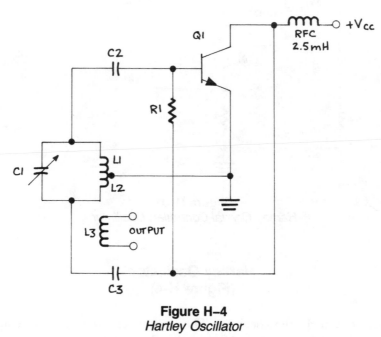

Figure H–4
Hartley Oscillator

Parts List

C1	Variable, value depends on frequency band*
C2	100 picofarads
C3	0.02 microfarad
L1, L2	Centertapped inductor, value depends on frequency band*
Q1	GE20, or equivalent general-purpose NPN transistor
R1	1 megohm

*Values may be calculated for frequency band required from:

$C = 1/39.48f^2L$

$L = 1/39.48f^2C$

where C = capacitance, L = inductance, and f = frequency

Hay Bridge
(Figure H-5)

This bridge is similar to the Maxwell bridge (q.v.), except that C1 and R2 are in series instead of parallel. The Hay bridge is useful for the measurement of inductors having fairly high values. The bridge is balanced by adjustment of R_s, and the unknown inductance and resistance calculated from:

$$L_X = R1C1R_s/1 + [(6.28f)^2C1^2R2^2]$$
$$R_X = R1R2R_s/1 + R2^2$$

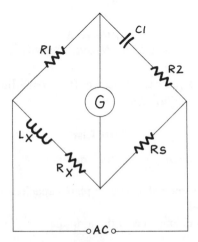

Figure H–5
Hay Bridge

Hazeltine Neutralization
(Figure H-6)

The fundamental amplifier circuit using a triode vacuum tube is practically identical with the circuit of a tuned-grid, tuned-plate oscillator (q.v.). Consequently it's hardly surprising that the amplifier also makes a good oscillator.

Oscillation occurs because the interelectrode capacitance of a triode tube permits positive feedback to occur from plate to grid, especially at radio frequencies. To prevent this from happening, a means must be provided for negative feedback to cancel the positive feedback. In this circuit a portion of the plate output signal is returned via C3 to the grid.

The primary of T2 is tapped at its center and grounded via C4 to give equal voltages at the ends of the tank circuit. The feedback connection is therefore made to the end that is opposite in polarity to the input signal at the grid. The neutralizing capacitor C3 has approximately

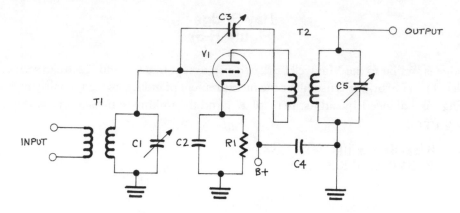

Figure H–6
Hazeltine Neutralization

the same capacitance as the grid-plate capacitance. It is then adjusted to match the neutralizing and feedback voltages. See *Amplifier Circuits*.

Parts List

C1, T1	Tuned to the frequency fed to the amplifier
C2	100 picofarads
C3	Approximately same value as grid-plate capacitance (1.6 picofarads)
C4	0.001 microfarad
C5, T2	Tuned to the frequency fed to the amplifier
R1	1.8 kilohms
V1	6C4

High-Pass Active Filter
(Figure H-7)

A high-pass filter passes higher frequencies but not lower ones, as in Figure H-7, where OF represents increasing frequency and OA increasing amplitude. The response of the filter is shown by the curve AF, which indicates that frequencies above f_c are passed readily, but that those below f_c are greatly attenuated. The range of frequencies above f_c is therefore called the passband, and the range lower than f_c is the stopband. The cutoff frequency f_c is defined as the frequency at which the amplitude A has fallen to 0.707A, or 3 dB down.

Figure H-8 gives the circuit of a commonly-used high-pass active filter. It will have a gain of 2 or 10 according to the values of the components selected. For a gain of 2:

R1 = 9 ohms
R2 = 56 ohms

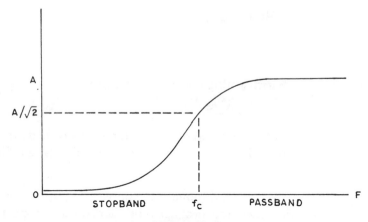

Figure H–7
High-Pass Active Filter Response Curve

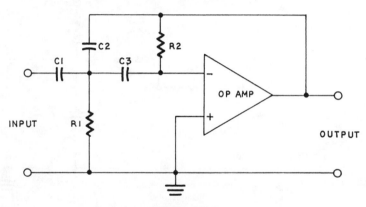

Figure H–8
High-Pass Active Filter Circuit

For a gain of 10 the values would be:

    R1 = 11 ohms
    R2 = 236 ohms

The value of C1 in microfarads is obtained by dividing f_c into 10. For example, if f_c is 1000, C1 would be $10/1000 = 0.01$ microfarad.

The value of C2 is $0.500 \times$ the value of C1 for a gain of 2, or $0.100 \times$ the value of C1 for a gain of 10. (The value of C1 is the same for either gain.)

The value of C3 is the same as the value of C1.

The operational amplifier IC used is the popular bipolar 741 type or the FET 536 type, or equivalent. See *Active Filter Circuits*.

High-Pass Passive Filter

See *Passive Filters*.

High-Voltage Power Supply
(Figure H-9)

The type of circuit used to provide a stable high-voltage supply for a cathode-ray tube is shown in Figures H-9 and H-10. Figure H-9 shows the high-voltage oscillator and Figure H-10 the high-voltage rectifier.

When power is first applied, the current through R11 charges C4 positive, so Q4 is forward biased. Its collector current increases, and a voltage is developed across the collector winding

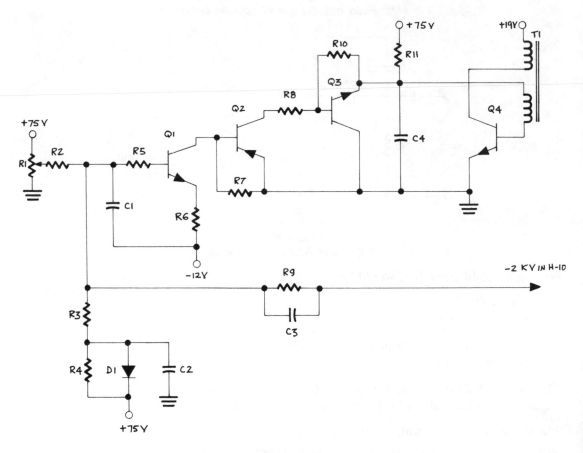

Figure H-9
High-Voltage Power Supply Oscillator

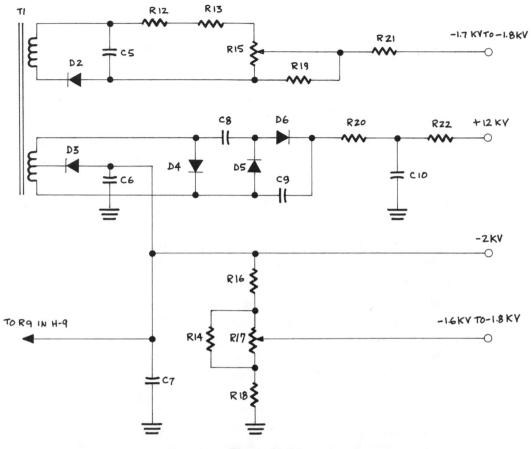

Figure H–10
High-Voltage Power Supply Rectifier

of T1. This produces a corresponding current increase in the feedback winding of T1, which further increases the voltage level on the base of Q4. When C4 is fully charged, the base current of Q4 stabilizes, so there is no changing current in T1's collector winding. Consequently no feedback voltage to Q4's base is induced in the feedback winding, and Q4's collector current begins to fall. This now produces a negative-going voltage on the base of Q4, which begins to turn the transistor off.

C4 slowly discharges to the negative potential on Q4's base. Once again the current flow through the collector winding ceases, and there is no feedback to the base of Q4. Then C4 begins to recharge through R11, and another cycle begins. The frequency of oscillation is between 40 and 50 kilohertz. This frequency is used because it requires smaller filter capacitors than a lower frequency. Larger capacitors would be impractical at the high voltages developed.

Q1, Q2, and Q3 comprise the voltage regulator circuit. A feedback from the output of the high-voltage rectifier circuit is applied to the base of Q1 via R9 and R5. This output voltage is negative. If it starts to go less negative (positive-going) it increases conduction in Q1, which in turn forward-biases Q2 to increase conduction in Q3. This raises the average voltage at Q3's emitter, which also appears on Q4's base. Q4's conduction increases, and a larger voltage is induced in the secondary of T1. This increases the negative output of the high-voltage rectifier so that the positive-going change is corrected.

Figure H-10 shows the high-voltage rectifier circuit. T1 has two high-voltage secondaries, which provide the negative and positive accelerating voltages, and the grid bias for the CRT.

Positive accelerating potential for the CRT anode is supplied by the voltage tripler D4, D5, and D6. This rectified voltage is filtered by R20, C10, and R23 to give a constant output of about 12 kilovolts.

The negative accelerating potential for the CRT cathode is supplied by the half-wave rectifier D3. The voltage output from this circuit is about −2 kilovolts (this is the supply from which the feedback sample is taken).

The half-wave rectifier D2 provides a negative voltage for the control grid of the CRT. R15 is provided for bias adjustment. The voltage divider R16, R17, and R18 provides voltage for focusing the CRT, and R17 is therefore the focus control. Other control voltages for the CRT (astigmatism, geometry, Y-axis alignment, and trace rotation) are provided from the low-voltage power supply. See *Power Supply Circuits*.

Parts List

C1, C4	0.01 microfarad
C2	2 microfarads
C3	0.001 microfarad
C5, C6, C7	0.015 microfarad, 2500 volts
C8, C9	500 picofarads, 10 kilovolts
C10	500 picofarads, 20 kilovolts
D1	1N4152
D2, D3	MPS−6521
D4, D5, D6	Silicon, 10 kilovolts, 5 milliamperes
Q1	2N2484
Q2	2N4122
Q3	2N3053
Q4	2N3055
R1	100 kilohms, variable
R2	4.3 megohms
R3	1 megohm

R4	200 kilohms
R5	10 kilohms
R6	1 kilohm
R7	470 kilohms
R8	100 ohms
R9	24 megohms
R10	100 kilohms
R11	30 kilohms
R12	10 kilohms
R13	50 megohms
R14	10 megohms
R15	2 megohms, variable
R16	13.2 megohms
R17	5 megohms, variable
R18	1.5 megohms
R19	7.5 megohms
R20, R21, R22	1 megohm

Hot Carrier Diode Mixer

See *Cowan Bridge Balanced Modulator.*

IF Amplifier

See AM, FM, or TV IF Amplifier.

IF Amplifier with Ceramic Filter
(Figure I-1)

This circuit illustrates the use of a ceramic band-pass filter between the mixer or converter stage of a radio receiver and an IF amplifier integrated circuit. The filter replaces the 455-kilohertz IF transformer, and has the advantage that it replaces all the IF transformers and needs no adjustment. See *Amplifier Circuits.*

Parts List

A1	LM172	C3	1 microfarad
C1	0.01 microfarad	C4	0.01 microfarad
C2	0.1 microfarad	FL1	Murata SF455D

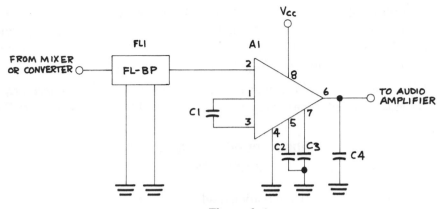

Figure I–1
IF Amplifier with Ceramic Filter

IIL or I²L NOR Gate
(Figure I-2)

The full name of this gate is "integrated injection logic." It is also called "merged transistor logic" (MTL) from the method of manufacture. Since it is found only in integrated circuits no parts list is provided.

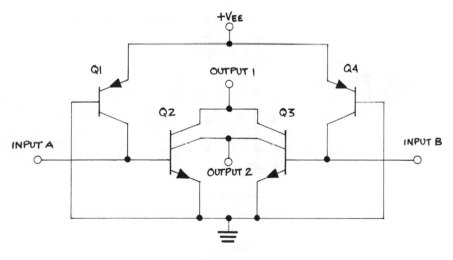

Figure I–2
IIL NOR Gate

Basically it operates in the same way as an RTL NOR gate (q.v.). However, input resistors are not used. Instead, there are two additional transistors Q1 and Q4, which provide drive currents for the bases of Q2 and Q3. Since Q1 and Q4 are PNP transistors, they inject an excess of minority carriers into the NPN transistors Q2 and Q3.

This is a very fast gate, and capable of high fan-out by the use of multiple-collector transistors, as shown. A logical 1 input at either A or B gives a logical 0 output. See *Logic Circuits*.

Infinite Impedance Detector
(Figure I-3)

In this circuit the triode tube is a combination diode rectifier and cathode follower (q.v.). The grid corresponds to the anode of a diode, and the rectifying action is the same as described for the crystal detector (q.v.). The DC voltage from rectified current flow through R1 charges C3 to the peak value of the rectified voltage on each pulse, and the capacitor retains enough charge between pulses so that the voltage across R1 is smoothed out. C3 therefore acts as a filter for the RF output of the rectifier, leaving a DC component that varies in the same way as the modulation on the original signal. C4, however, transmits only the variations in the voltage, so the final output signal is AC. This is applied across the volume control R3. See *Demodulation Circuits*.

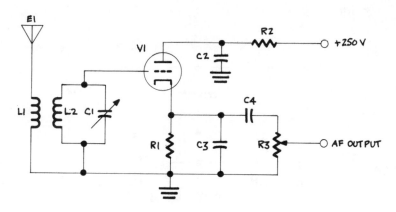

Figure I–3
Infinite Impedance Detector

Parts List

C1   28 MHz: 15 picofarads, variable
C2   0.001 to 0.002 microfarad
C3   100 to 250 picofarads
C4   0.1 microfarad
L1   28 MHz: 4 turns of No. 30 enameled copper wire*

L2   28 MHz: 8 turns of No. 30 enameled copper wire*
R1   150 kilohms
R2   25 kilohms
R3   250 kilohms, variable ("volume control")
V1   6C4

*Both coils should be close-wound on a Millen 74001 1/2-inch diameter coil form with tuning slug (or equivalent).

Instant-On Radio or TV Power Control
(Figure I-4)

This circuit is suitable for transformerless power supplies in which the heaters of the tubes are in series across the power line. The half-wave rectifier is shown as a vacuum-tube diode V5, but the principle is the same for solid-state rectifiers.

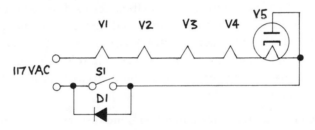

Figure I–4
Instant-On Radio or TV Power Control

With S1 open, and D1 (5 amperes, 200 volts PIV) connected across it with the polarity shown, only the negative half-cycles of the power-line voltage are applied to V5's anode. Consequently the rectifier cannot conduct, and no B + supply is provided for the receiver's tubes. However, these negative pulses do flow through the tube heaters, keeping them warm until they are required to operate. Therefore, when S1 is closed and positive half-cycles are applied to V5's anode, the receiver plays immediately, as no warm-up time is required. See *Power Supply Circuits*.

INTEGRATED CIRCUITS

Integrated circuits were introduced in 1961, and since then have been the fastest growing segment of electronic component technology. An IC consists of a combination of transistors, resistors, and diodes that are formed on the surface of a substrate, generally of silicon. These components are of such microscopic size that this branch of electronics is termed microelectronics. Circuit functions performed by ICs fall into one or the other of two main categories: digital and analog.

Digital circuits. The majority of these are for logic circuits (q.v.), such as those used in computers. They range from small-scale ICs providing simple gating functions, through more complex single-function devices called medium-scale integrated circuits (MSI), to even larger system functions called large-scale integrated circuits (LSI), and very large-scale integrated circuits (VLSI). Among the latter are microprocessors, memories, and other extremely complex devices that have made possible the modern boom in personal computers, TV games, "smart" typewriters, and the like.

Analog circuits. Also called linear ICs, these include operational amplifiers (q.v.), differential amplifiers, microwave devices, and many others. Since integration has resulted in a 100-fold cost reduction in the manufacture of such circuits, it is now possible to use configurations such as the phase-locked loop that were previously uneconomical.

The fabrication of these devices involves the photo-etching of the surface of the silicon "chip," along with the implantation, deposition, and diffusion of various impurities, insulating layers, and metal conductors. It is important to recognize that these compact devices of necessity provide only relatively small insulation areas and thicknesses. For instance, the insulated gate of a MOSFET is typically 0.1 micrometer thick, and therefore is susceptible to damage by electrostatic discharge. This can happen if such a device is merely handled, since the handler's body capacitance can build up a considerable charge. Therefore, prior to assembly, all leads should be kept shorted together by insertion into conductive material such as "ECCOSORB LD26" (a product of Emerson and Cumming, Inc.).

When being assembled on a board, the assembler's hand should be grounded, together with his or her soldering iron tip. It goes without saying that MOS devices should never be inserted or removed from equipment with the power on.

Many modern ICs now provide protection against electrostatic effects by built-in circuitry. Usually this consists of connecting diodes between each transistor's insulated gate and source. Even so, the same precautions should be taken, to be on the safe side.

The leads of in-line packages can be bent, if necessary, but they are not flexible in any general sense, not are they sufficiently rigid for unrestrained wire-wrapping. Great care is required to ensure that the lead is not loosened where it enters the plastic case, and that it is not bent repeatedly, or it will break. Soldering to the leads can be done safely, if the temperature does not exceed 275°C, and is not applied for longer than five seconds, but a much better procedure is to use DIP sockets, if there is enough room.

Integrated circuits are great time-savers, since each is a functional block rather than a single component in such a block. Some popular ICs that are generally available in consumer electronics outlets are listed below:

DESCRIPTION	CMOS	TTL	SCHOTTKY
Quad 2-input NAND gate	4011	7400	74LS00
Quad 2-input NOR gate	4001	7402	74LS02
Dual D flip-flop	4013	7474	74LS74
Decade counter	4017	7490	74LS90
Dual J-K flip-flop	4027	7473	
Quad latch	4042	7475	74LS75
Hex inverter	4049	7404	74LS04
8-bit shift register			74LS164

	TYPE
Static RAM (1024 bits)	2102L
Static RAM (4096 bits)	2114L
Dynamic RAM (16,384 bits)	4116
Timer	555
Voltage regulator	LM723
Operational amplifier	741C
IF amplifier	MC1350
Video detector	MC1330
Phase-locked loop	LM565
Audio amplifier	LM383
Sound generator	SN76477
Sound detector	MC1358

Examples of Employment of Integrated Circuits

See the following list for examples of employment of integrated circuits described elsewhere in the encyclopedia:

AM Radio Using IC
Audio Amplifier with IC
Audio Power Output Stage with IC Driver
Audio Preamplifier with IC
Cascode JFET Preamplifier
Electronic Oven Control
Frequency-Modulated IF Amplifier with IC
IC Audio-Frequency Amplifier
IC Audio-Frequency Preamplifier

IC Bridged-T Oscillator
IC FM IF Amplifier
IC Function Generator
IC Intercom
IC Phase-Shift Oscillator
IC Power Amplifier
IC Stereo Preamplifier
LOGIC CIRCUITS
TV IF Amplifier with IC

IC Audio-Frequency Amplifier

See *Audio Amplifier with Integrated Circuit.*

IC Audio-Frequency Preamplifier

See *Audio Preamplifier with Integrated Circuit.*

IC Bridged-T Oscillator
(Figure I-5)

This circuit employs an operational amplifier with positive feedback to the non-inverting input. Negative feedback is also applied to the inverting input via a bridged-T network. This bridged-T network provides maximum attenuation of the negative feedback at its null frequency, so that the circuit oscillates at this frequency. The lamp resistance increases as the feedback increases. This limiting action improves the stability of the circuit. See *Oscillator and Signal Generator Circuits; Integrated Circuits.*

Parts List

A1	MC1456CG
C1, C2	0.01 microfarad
C3, C4	10 microfarads
DS1	Sylvania 120MB
R1	30 kilohms
R2	6.8 kilohms
R3	300 kilohms
R4	1 kilohm
R5	3 megohms
R6	1 kilohm, variable
R7	10 kilohms

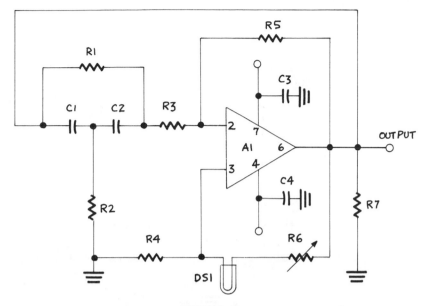

Figure I–5
IC Bridged-T Oscillator

IC FM IF Amplifer

See *FM IF Amplifier with IC.*

IC Function Generator
(Figure I-6)

This circuit can provide two outputs: a triangular wave and a square wave. S1 selects frequency bands and R3 tunes the required frequency in the band selected. Nearly complete coverage is obtained from 20 hertz to 1 megahertz. See *Oscillator and Signal Generator Circuits; Integrated Circuits.*

Parts List

A1   NE566T (Signetics)
C1   0.001 microfarad
C2   1 microfarad
C3   0.1 microfarad
C4   0.01 microfarad
C5   0.001 microfarad
C6   0.0001 microfarad

R1 2.2 kilohms
R2 10 kilohms
R3 18 kilohms
R4 2 kilohms
S1 Single-pole, five-position rotary switch

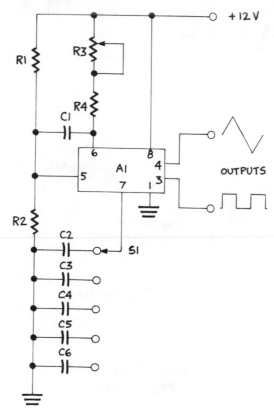

Figure I–6
IC Function Generator

IC IF Amplifier

See *AM, FM,* or *TV IF Amplifier.*

IC Intercom
(Figure I-7)

This circuit is a simple way to build an intercom unit. The two loudspeakers are used alternately as speaker and microphone according to the setting of the "push-to-talk" switch S1. When the "master" station turns on the power, using S2, and depresses S1, LS2 becomes his microphone and LS1 the "slave" speaker. On releasing S1, LS1 becomes the microphone, allowing the master to hear. See *Amplifier Circuits; Integrated Circuits.*

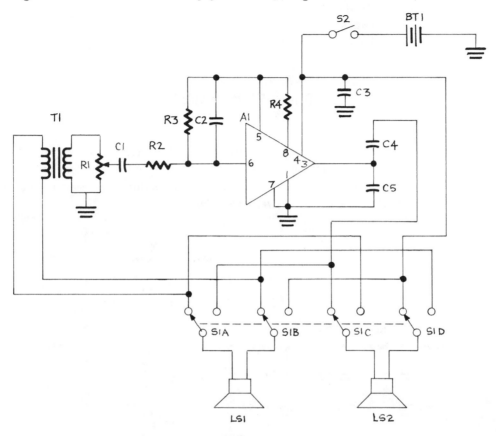

Figure I–7
IC Intercom

Parts List

A1	MC1306P or equivalent
BT1	9-volt rect. battery
C1	0.01 microfarad
C2	12 picofarads
C3	1 microfarad
C4	200 microfarads
C5	0.05 microfarad
LS1, LS2	Speakers, 4 ohms, 2 to 3 inches diameter
R1	5 kilohms, variable ("volume control")
R2	10 kilohms

R3	1 megohm
R4	1 kilohm
S1	Four-pole, double-throw, momentary pushbutton switch
S2	On-off switch, any type
T1	Microphone transformer to transform voice-coil impedance of 4 ohms to 10–15 kilohms

IC Phase-Shift Oscillator
(Figure I-8)

In this circuit feedback from the operational amplifier is connected to its inverting input via an RC network that gives 180 degrees of phase shift. Consequently, the feedback signal is inverted and applies a positive feedback signal to the input, the condition necessary for oscillation. See *Oscillator and Signal Generator Circuits; Integrated Circuits.*

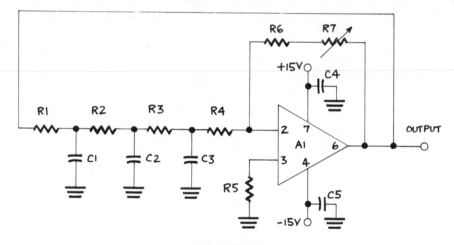

Figure I–8
IC Phase-Shift Oscillator

Parts List

A1	MC1456CG
C1, C2, C3	0.033 microfarad
C4, C5	10 microfarads
R1, R2, R3	10 kilohms
R4	100 kilohms
R5	130 kilohms
R6	2 megohms
R7	2 megohms, variable (''gain adjust'')

IC Power Amplifier

See *Audio Power Output Stage with IC Driver.*

IC Quadrature FM Detector
(Figure I-9)

Although there are ICs available (e.g., MC1358) that include on the same chip an IF amplifier, limiter, FM detector, and audio driver, this circuit shows how a quadrature FM detector can be built using a simple IC amplifier. See *Demodulation Circuits; Integrated Circuits.*

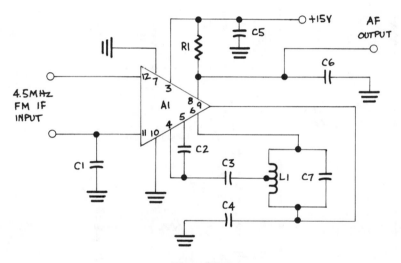

Figure I–9
IC Quadrature FM Detector

Parts List

A1	PA189 (GE)
C1, C2, C3, C4	0.005 microfarad
C5	0.05 microfarad
C6	0.001 microfarad
C7	15 picofarads
L1	83 microhenries (4.5 MHz quadrature coil)
R1	470 ohms
T1	4.5–MHz IF transformer

IC Stereo Preamplifier
(Figure I-10)

This circuit shows one channel, using one half of the IC (the other half is used by the other channel). The preamplifier has low noise and high gain, and would be suitable for tape input. See *Amplifier Circuits; Integrated Circuits.*

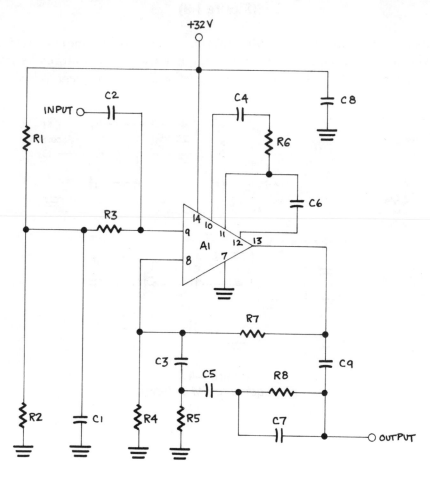

Figure I–10
IC Stereo Preamplifier

Parts List

A1	Half of a μA739 (Fairchild)
C1, C2,	5 microfarads
C3, C9	
C4, C6	0.005 microfarad

C5	0.0027 microfarad
C7	150 picofarads
C8	0.01 microfarad
R1	470 kilohms
R2	47 kilohms
R3	150 kilohms
R4	270 kilohms
R5	1 kilohm
R6	10 ohms
R7	1 megohm
R8	100 kilohms

Integrator
(Figure I-11)

This circuit is used to integrate the six vertical sync pulses in the television signal so that they become one large pulse with an amplitude high enough to trigger the vertical oscillator. When the vertical sync pulses arrive at the integrating circuit, they charge the integrating capacitors slowly because of the long time constant. Between charges the very long time constant of the integrator prevents any appreciable discharge of the integrating capacitors. This means the integrating capacitors charge in steps, rising to an appreciable level during the interval of the vertical sync pulses until the amplitude of the charge reaches the triggering level. Integrator circuits vary a bit. The one illustrated would probably be provided in the form of a printed circuit.

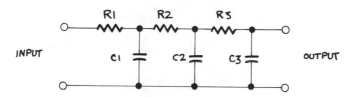

INPUT TV SYNC PULSES:
HORIZONTAL (NARROW), VERTICAL (BROAD)
OUTPUT INTEGRATED VERTICAL
TRIGGER PULSE (X—X)

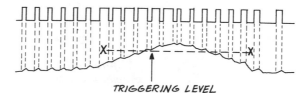

TRIGGERING LEVEL

Figure I-11
Integrator

Parts List

C1	0.002 microfarad
C2, C3	0.005 microfarad
R1	22 kilohms
R2, R3	8.2 kilohms

Intermediate-Frequency Amplifier

See *AM, FM,* or *TV IF Amplifier.*

Inverter
(Figure I-12)

Inversion of signal polarity is often required and, for this, advantage is taken of the property of the common-emitter or common-source amplifier to give an output of opposite phase to its input. Figure I-12 shows a bipolar transistor circuit, while Figure I-13 shows a CMOS circuit that avoids the use of resistors, making it suitable for use in ICs. No parts list is given for either circuit, as they are used only in integrated circuits. See *Logic Circuits.*

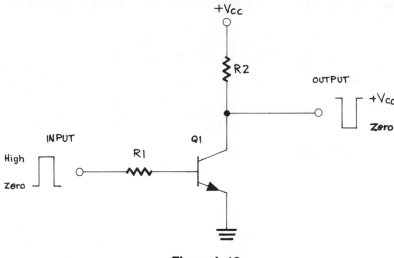

Figure I–12
Bipolar Inverter

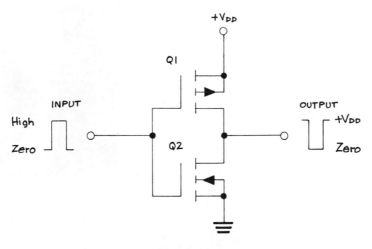

Figure I–13
CMOS Inverter

Kelvin Bridge (Figure K-1)

The Kelvin bridge is used for measuring very low resistances. A Wheatstone bridge (q.v.) with six dials has total **switch zero resistance,** or "contact resistance," of 0.0035 to 0.004 ohm, which can have considerable effect on measurements lower than 0.1 ohm.

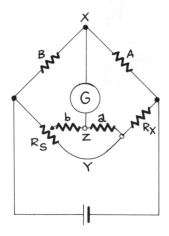

Figure K–1
Kelvin Bridge

Even if the switch zero resistance is negligible, there is another difficulty. Suppose you want to measure a resistance of 0.001 ohm, and you require the value to four places. The bridge has a four-dial rheostat, the highest decade being X1000. The ratio between R_S and R_X is, therefore, 1000 : 0.001, or $10^6 : 1$. To balance this, the A/B ratio will have to be $10^6 : 1$ also. You can see that almost the entire battery voltage will be dropped across R_S and B. The 1000-ohm resistor in the rheostat arm R_S, with a wattage of 0.5 watt, cannot take more than 22.36 milliamperes, which restricts the battery voltage to 22.36 volts. Since the voltage at Y will be approximately one millionth of this, it will give no indication at all on the galvanometer.

By replacing the rheostat arm R_S with a slide wire or bar of Manganin with a sliding contact, the ratio between the standard and unknown resistances can be brought into better proportion. However, the current increase makes the voltage dropped across the resistance of the sliding contact larger, so the gain from the increased sensitivity is counterbalanced by the increased switch zero resistance.

The Kelvin bridge gets around this difficulty by adding two additional arms, a and b. As long as the ratio a/b equals A/B, R_X will still equal R_S A/B, but because the resistance values of a and b are so much higher than R_S and R_X practically all the current flows through R_S – Y – R_X. The current through b – Z – a is negligible and, therefore, the voltage dropped across the sliding contact is no longer a problem.

The Wheatstone and Kelvin bridges are able to handle resistance measurements from 0.001 ohm to over 10 megohms. Because of their similarity, they can be combined in one instrument. See *Test and Indicating Instruments*.

Key-Click Filter (Figure K-2)

Nearly all keying installations require the type of filter shown here to minimize interference caused by the spark (often very minute) at the key contacts. Suitable parts values are best found by experiment, although those given in the following parts list represent good starting points.

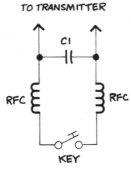

Figure K–2
Key-Click Filter

Parts List

C1   0.001 microfarad, with voltage rating at least as high as voltage across key when key is up
RFC   RF chokes, 2.5 millihenries, able to carry key current

Keyed AGC (Figure K-3)

This circuit illustrates the principles of keyed or gated automatic gain control. Parts values are given, although you are not likely to want to build this circuit.

The keyer tube is a pentode, with its cathode maintained at + 125 volts. The grid is biased to hold the tube at cutoff. Sync pulses from the first video amplifier act on the grid to overcome most or all of the bias.

V1's plate is not connected to B + . Consequently, there is no way this tube will conduct under such a condition, especially with its cathode at a high positive voltage, regardless of what is happening on the grid. However, there is a connection from the plate via C3 to the horizontal output transformer, so it receives positive pulses at the horizontal line frequency. The voltage of these pulses is considerably higher than that on the cathode. Each time a pulse is received the tube is able to conduct, and as this coincides with the receipt of a sync pulse on the grid, the tube conducts in accordance with the amplitude of the latter.

Pulses of keyer conduction current flowing away from V1's plate impart a negative charge to C2. This negative charge is the AGC voltage, which increases with stronger received signals, and decreases with weaker signals. See *Amplifier Circuits*.

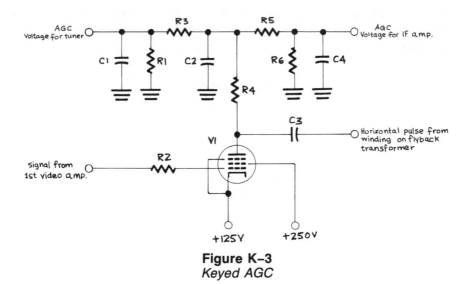

Figure K–3
Keyed AGC

Parts List

C1	0.05 microfarad
C2	0.005 microfarad
C3	1000 picofarads (1000 volts)
C4	0.5 microfarad
R1	4.7 megohms
R2	33 kilohms
R3	330 kilohms
R4	100 kilohms
R5	150 kilohms
R6	50 kilohms
V1	6AU6 or equivalent

Laboratory Power Supply (Figure L-1)

The construction of this power supply is described on page 345. The following, therefore, is a circuit description and list of parts.

T1, D1, D2, C1, and C2 comprise a voltage doubler (q.v.). Its unloaded output is approximately 35 volts DC. Current through a load connected to the output jacks J1 and J2 is regulated by the Darlington pair consisting of Q1 and Q2. Q1 and Q2 are controlled by the error amplifier A1. Its inverting input is connected to the positive output jack, and its non-inverting input is connected to the negative output jack via R3. Varying R3 gives output voltages at the output jacks ranging from 0 to 15 volts DC.

The error amplifier has its own power supply consisting of T2, D3, D4, C3, and C4. This is another voltage doubler with an unloaded output of approximately 17 volts DC, that supplies A1 with a $+V_{CC}$ of $+8.4$ volts and a $-V_{EE}$ of -8.4 volts. See *Power Supply Circuits*.

Parts List

A1	Operational amplifier 741 or equivalent
C1, C2	3300 microfarads, 35 volts
C3, C4	220 microfarads, 16 volts
D1, D2, D3, D4	Silicon rectifiers, 1N4000 series, PIV 50
D5	Zener diode, 1N4735, 6.2 volts

J1, J2	Panel-mounted jacks or binding posts, red and black
M1	0-1 ampere current meter
M2	0-15 volt meter
Q1, Q2	Darlington pair, ECG261, or equivalent
R1	470 ohms, 1/2 watt
R2	3.3 kilohms, 1/2 watt
R3	10 kilohms, variable (linear potentiometer)
S1	On-off switch (mounted on rear of R3)
T1	Power transformer, 120/12.6 volts, 1.2 amperes
T2	Power transformer, 120/6.3 volts, 1.2 amperes

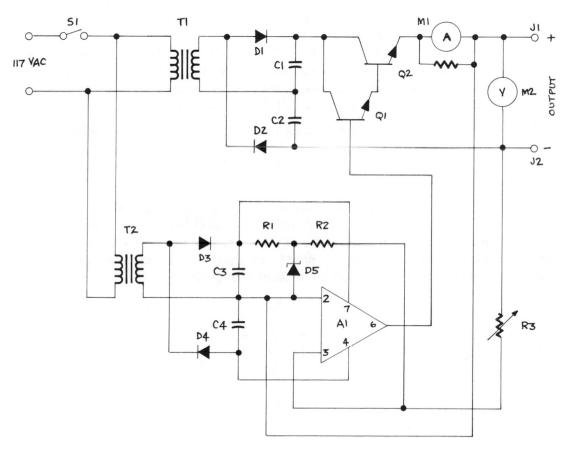

Figure L–1
Laboratory Power Supply

Level Control

See *Automatic Level Control.*

Light Dimmer (Figure L-2)

This simple light-dimmer circuit contains a diac, triac, and RC charge-control network. A diac is a two-terminal AC switch that is changed from the nonconducting to the conducting state by an appropriate voltage of either polarity. The triac is a three-terminal AC switch that is changed from the nonconducting to the conducting state when a positive or negative voltage is applied to the gate terminal. The diac is used to trigger the triac.

During the beginning of each half-cycle of the AC line voltage the triac is in the off state. As a result, the entire line voltage appears across the triac, and none across the load (e.g., a lamp.) Because the triac is in parallel with R2 and C2, the voltage across the triac drives current through R2 to charge C2. When the capacitor voltage reaches about 35 volts (this is called the breakover voltage, V_{BO}, of the diac), the diac's resistance decreases sharply, and the capacitor discharges through the diac and the triac gate, turning on the triac. At this point, the line voltage is transferred from the triac to the load for the remainder of that half-cycle. This sequence of events is repeated for every half-cycle of either polarity.

If R2's resistance is reduced, C2 charges more rapidly, and V_{BO} is reached earlier in the cycle. This increases the power applied to the load, so the light is brighter. Conversely, if the resistance of R2 is increased, V_{BO} is reached later in the half-cycle, less power is applied to the load, and the light is dimmer.

The rapid switching of the triac could result in radio noise affecting the AM broadcast and short-wave bands (but not FM or TV); therefore, a suppression network consisting of C1 and L1 is included in the circuit. It is also important to note that a triac dissipates heat at the rate of about 1 watt per ampere, so it must be mounted on an effective heat sink (the face plate and wall box serve well enough up to about 6 amperes).

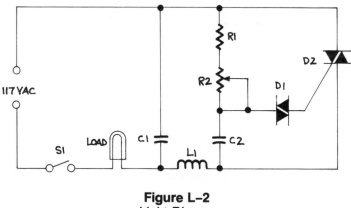

Figure L–2
Light Dimmer

Parts List

C1, C2	0.05 microfarad, 100 volts
D1	Diac, D3202U, or equivalent
D2	Triac, T2800B, or equivalent
L1	RF choke, 100 microhenries
R1	3300 ohms, 1/2 watt
R2	250 kilohms, 1/2 watt
S1	On-off switch (can be mounted on rear of R2)

Limiter

See *Clipper.*

Linear Amplifier (Figure L-3)

This circuit is that of a typical Class-B linear RF amplifier. A pentode tube is used, although a triode with neutralization (q.v.) or a tetrode could also be used. The modulated carrier is coupled to the grid of V1 via the input resonant circuit and C3. A fixed bias voltage is applied to the grid via R1. This bias is set at approximately the cutoff point of V1.

Since the negative half-cycles of the input signal extend into the cutoff region, the tube can conduct only on the positive half-cycles. The plate current, therefore, flows in pulses that vary in amplitude according to the modulation. However, these shock-excite the output resonant circuit into oscillation so that the missing half-cycles are reintroduced (the "flywheel effect"). See *Amplifier Circuits.*

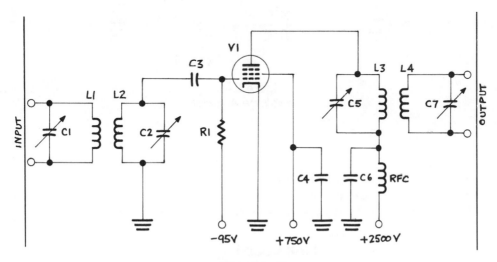

Figure L-3
Linear Amplifier

Parts List

C1, C2,   Variable capacitors for tuning the resonant circuits (values depend on frequency band
C5, C7   used)

C3, C4,   250 picofarads
C6

L1, L2,   Inductors with values as required for the frequency band used
L3, L4

R1   50 kilohms

RFC   RF choke, 10 microhenries, or according to frequency requirements

V1   813 or similar

Line Voltage Adjuster (Figure L-4)

This circuit requires a power transformer with a center-tapped secondary. Suppose T1 has a 12.6 volt secondary with a center tap. The line voltage will induce 6.3 volts in each half of the winding. Assuming a line voltage of 120 volts, when S1 is in position 1 the voltage at J1 will be 120 + 6.3 = 126.3 volts; in position 2 the voltage at J1 will be 120 + 0 = 120 volts; and in position 3 the voltage at J1 will be 120 − 6.3 = 113.7 volts.

This is not as flexible as a variable autotransformer, but is useful where the line voltage varies considerably at different times of the day, and is definitely less expensive. See *Power Supply Circuits*.

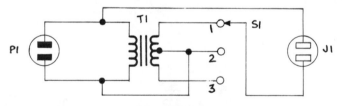

Figure L–4
Line-Voltage Adjuster

Parts List

J1   AC socket
P1   Two-prong AC plug, with line cord
S1   Three-position switch, any type
T1   Power transformer, 120/12.6 volts, with center-tapped secondary

Link Coupling

See *Loop Antenna, Inductively Coupled.*

Logic Circuits

A logic circuit is one in which discrete voltage or current levels are used to represent the binary digits 0 and 1 to carry out one of the logic functions. Such a circuit therefore is also called a digital circuit.

Binary digits are called **bits** for short, and a set of eight is called a **byte**. Since there are only two bits, 0 and 1, it is simple to use the absence or presence of some voltage to indicate them. The voltage value will depend upon the type of logic circuit, but in the most widely used bipolar logic circuit a 0 is no voltage (actually a very small voltage), and a 1 is approximately +5 volts.

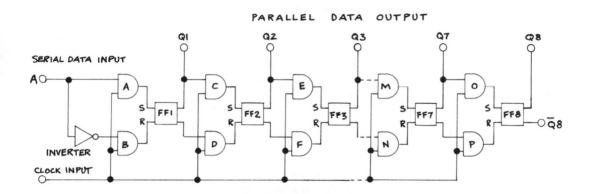

Figure L–5
Eight-Bit Shift Register

The circuits that manipulate these bits consist mainly of storage devices and switches. A set of eight flip-flops (q.v.) can store a byte and is called a register. Each flip-flop will have an output voltage of 0 or +5 which it will retain until reset. Thus, the eight-bit shift register in Figure L-5 would store the binary number 01101001 as 0, +5, +5, 0, +5, 0, 0, +5 volts at the output terminals of its eight flip-flops. These are RS flip-flops. An RS flip-flop is made up of a number of **gates**, as shown in Figure L-6. Before explaining how this flip-flop works we need to talk about gates.

Two switches placed in series with a battery and a lamp form an **AND** gate (Figure L-7). If both switches are closed the lamp lights. If either, or both, switches are open the lamp does not light. "A and B equal C," means "when A *and* B are both closed, the lamp C lights."

If you place the switches in parallel (Figure L-8), closing either switch will light the lamp. This time "A or B equals C," meaning "when either A *or* B is closed, the lamp C lights." This arrangement is therefore an **OR** gate.

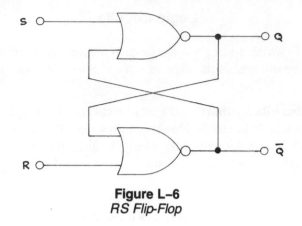

Figure L–6
RS Flip-Flop

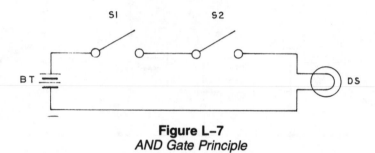

Figure L–7
AND Gate Principle

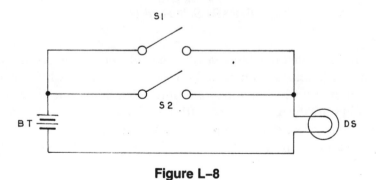

Figure L–8
OR Gate Principle

A third type of gate is called an **inverter**. As you can see from Figure L-9, this is no more than an ordinary common-emitter amplifier circuit (q.v.), only here it is not being used as an amplifier. One of the properties of a common-emitter amplifier is that the output is of opposite polarity to the input, so if you want to invert a bit (change it from a 0 to a 1, or vice versa) you apply it to an inverter.

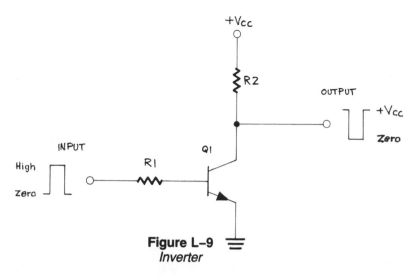

Figure L-9
Inverter

Gates are usually shown in diagrams by graphic symbols, as in Figure L-10. The triangle for the inverter is the general symbol for an amplifier, with a little circle added to indicate inversion. By adding this circle to the symbols for AND and OR gates, they are turned into NAND and NOR gates, in which the output bit is inverted. The functions of these gates with various inputs are summarized in **truth tables**. A truth table for an AND gate is shown in Figure L-11. The condition of each switch is shown under A and B (0 = "open," 1 = "closed"), and the result at the output is shown under C (0 = "lamp is not lit," 1 = "lamp is lit"). The truth table for a NAND gate would be the same, except that the output states under C would be reversed.

Returning to the RS flip-flop (Figure L-6), you can now see how it works. A 1 at the S input and a 0 at the R input will cause the NOR gate A to switch, giving a 0 output, while the NOR gate B will give a 1 output, because both inputs will be 0. When these flip-flops are cascaded (with AND gates) as in Figure L-5, a data bit applied to AND gate A simultaneously with a clock pulse will result in a 1 bit being applied to the S input of flip-flop 1. At the same time an inverted data bit (0) is applied to AND gate B, so a 0 bit is applied to the R input of the flip-flop. The output at Q1 will therefore be a 1. On the next clock pulse this 1 will be transmitted to the S input of flip-flop 2, and on succeeding clock pulses, to flip-flop 3, 4, and so on.

Meanwhile, succeeding data bits are arriving at A. If they are 1's, the same sequence of operations takes place, but if a 0 arrives, a 0 will be applied to the S input of flip-flop 1, and a 1 to its R input. This will cause a reversal of the outputs, and the Q1 output will be a 0, which will also progress through the register in the same way. When all eight flip-flops have received data bits, one byte will have been stored. It can be recovered in serial or parallel form according to the way the output is connected.

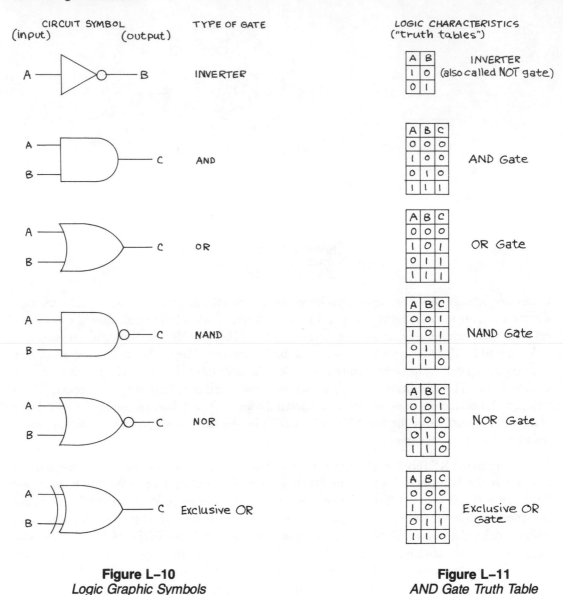

CIRCUIT SYMBOL (input) (output)	TYPE OF GATE	LOGIC CHARACTERISTICS ("truth tables")

INVERTER

A	B	
1	0	INVERTER (also called NOT gate)
0	1	

AND

A	B	C
0	0	0
1	0	0
0	1	0
1	1	1

AND Gate

OR

A	B	C
0	0	0
1	0	1
0	1	1
1	1	1

OR Gate

NAND

A	B	C
0	0	1
1	0	1
0	1	1
1	1	0

NAND Gate

NOR

A	B	C
0	0	1
1	0	0
0	1	0
1	1	0

NOR Gate

Exclusive OR

A	B	C
0	0	0
1	0	1
0	1	1
1	1	0

Exclusive OR Gate

Figure L–10
Logic Graphic Symbols

Figure L–11
AND Gate Truth Table

We've been talking about switches all this time, but obviously they're not toggle switches. Transistors are used in simple circuits such as that in Figure L-12. In this example of a NOR gate, the two transistor switches are connected in parallel between the V_{CC} supply and the low side of the circuit, which can be assumed to be at zero potential. In the absence of any 1 bit on the base of either Q1 or Q2, the transistor switches are nonconducting, or open. Since no current flows through the load resistor there can be no voltage drop across it, so the 5 volts

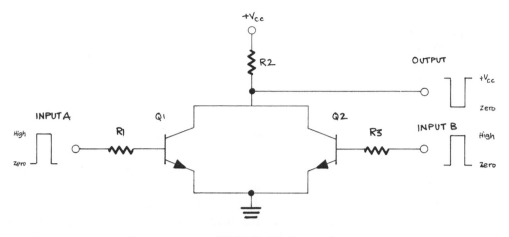

Figure L–12
Transistor Switches in a NOR Gate

of the power supply appears on the collectors of the transistors, and therefore at the output C. In other words, with both inputs 0, the output is 1.

When a 1 bit appears at either base, that transistor turns on and becomes a very low resistance. Practically all the supply voltage is now dropped across the load resistor, and the voltage at the output drops to 0. In other words, a 1 input on either, or both, bases, gives a 0 output.

All transistor switches work in much the same way, although some work much faster than others, or use more power, or can output to several different circuits ("fan-out"). There are also various power supply voltages (MOS devices generally use higher levels). Particular devices are described elsewhere in the encyclopedia.

Other Logic Circuits

See the following list for other logic circuits described elsewhere in the encyclopedia:

AND gate
CML NOR gate
CML OR gate
CTL OR gate
DCTL NAND gate
DCTL NOR gate
DTL NAND gate
Flip-flop
IIL or I^2L gate
Inverter
NAND gate

NOR gate
OR gate
RCTL NOR gate
RTL NOR gate
Scaling Inverter with Operational Amplifier
TTL NAND gate

Loop Antenna (Figure L-13)

Before the ferrite-core antenna was introduced, portable radios used an antenna consisting of a number of turns of insulated wire glued spirally on a nonconducting surface, such as the rear panel of the receiver. Before that, when radios were not generally portable, many apartment dwellers and others living in places where outside antennas were not allowed, or were not feasible, had indoor antennas consisting of many turns of wire wound around a wooden frame, which stood on a table or the floor, and could be quite large. The "effective height" (h_e) in meters of such an antenna is given by:

$$h_e = 6.28nAf$$

where n is the number of turns, A the mean area in square meters per turn, and f the frequency in hertz.

Both types were markedly directional and had to be turned for maximum sensitivity until the broadcast station was at right angles to the horizontal axis of the antenna. This property is still used in the modern loop antenna widely employed by aircraft in the Airborne Direction Finder (ADF) system.

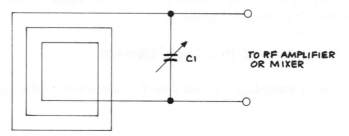

Figure L–13
Loop Antenna

This antenna consists of a ring of many turns of insulated wire, enclosed in a tube to protect it from the weather, and rotatable from inside the aircraft by means of a crank handle. As it rotates, pointers on a dial show the directions of maximum and minimum sensitivity in degrees of azimuth with respect to the fore-and-aft axis of the aircraft. The navigator tunes the radio to receive a broadcast station, or a Non-Directional Beacon (NDB) that operates in the 200 to 415 kilohertz band, and rotates the antenna for a null (direction of least sensitivity, which is sharper than a maximum, so is more generally used). The direction indicated by the null pointer can then be combined with the heading of the aircraft to give the bearing of the station at that moment.

While not the most accurate bearing measurement system, it is popular because it will work with almost any type of station in the 200 to 1600 kilohertz range. See *Antenna Circuits.*

Loop Antenna Coupling (Figure L-14)

The sensitivity of a radio receiver employing a loop antenna or ferrite-core antenna ("loopstick") can be greatly increased by using an external antenna and earth ground. The external antenna and ground are connected to a flat loop antenna (see *Loop Antenna*), which is placed close to the radio so that energy is transferred inductively to the radio's internal antenna. See *Antenna Circuits.*

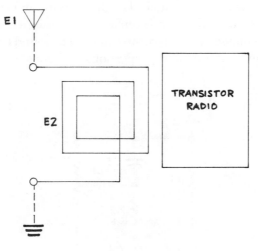

Figure L–14
Loop Antenna Coupling

Loop Antenna, Inductively Coupled (Figure L-15)

A loop antenna with only two or three turns of wire, provided that they form a large enough rectangle, may be used if it is inductively coupled to the receiver's input via an RF transformer, whose secondary is tuned by a capacitor to be resonant at the frequency. See *Antenna Circuits.*

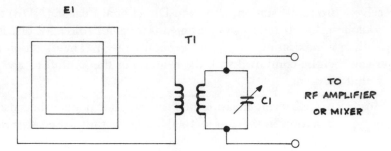

Figure L–15
Inductively-Coupled Loop Antenna

Loudness Control (Figure L-16)

As the output of an amplifier is reduced the apparent loudness of the low notes will be reduced more rapidly than the middle frequencies. In fact, the intensity of the sound must be comparatively high to hear low notes at all. For this reason, an ordinary volume control is not used in a high-fidelity amplifier, but is replaced by a loudness control which boosts the bass (and also the high frequencies) as it is turned down. The sound level at various frequencies is thus adjusted to maintain the desired proportions. See *Amplifier Circuits*.

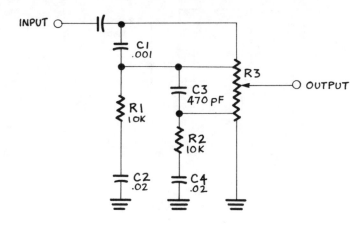

Figure L–16
Loudness Control

Parts List

C1	0.001 microfarad
C2, C4	0.02 microfarad

C3   470 picofarads
R1, R2   10 kilohms
R3   500 kilohms, variable, audio taper, tapped

Low-Frequency Attenuator

See *High-Pass Filter.*

Low-Impedance Frequency Doubler (Figure L-17)

This circuit is designed to receive a 21-megahertz input and convert it to a 42-megahertz output. The pi-network in the input is tuned to be resonant at 21 megahertz, and the double-tuned coupling transformer in the output is resonant at 42 megahertz. The coupling of the windings of this transformer is critical: too close will result in the passing of spurious harmonics, and too loose will not give an adequate output. Both input and output impedances are low. See *Frequency Conversion Circuits.*

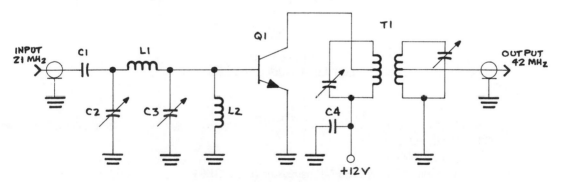

Figure L-17
Low-Impedance Frequency Doubler

Parts List

C1, C4   1000 picofarads
C2   170–780 picofarads, variable
C3   100–580 picofarads, variable
L1   0.21 microhenries
L2   27 microhenries (RF choke)
Q1   2N743 or equivalent
T1   Double-tuned coupling transformer for 42 MHz

Low-Level Audio Amplifier (Figure L-18)

This is a general-purpose audio amplifier that can be used for a variety of applications. I_C is 1 milliampere and is stabilized by degenerative feedback provided by R3. This resistor is bypassed by C2 to prevent AC degeneration. The circuit elements place Q1's operating point at about half the supply voltage, so the output voltage can swing between zero and − 12 volts. See *Amplifier Circuits*.

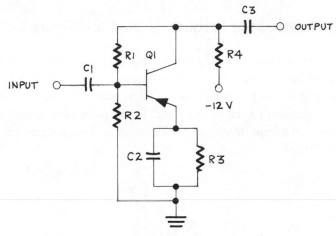

Figure L–18
Low-Level Audio Amplifier

Parts List

C1, C3   12 microfarads
C2   100 microfarads
Q1   Any general-purpose PNP transistor (e.g., GE84)
R1   56 kilohms
R2   15 kilohms
R3   1.5 kilohms
R4   4.7 kilohms

Low-Pass Active Filter

See *Active Filter Circuits*.

Low-Pass Passive Filter

See *Passive Filter Circuits*.

L-Type LC Filter

See *Passive Filter Circuits*.

L-Type RC Filter

See *Passive Filter Circuits*.

Magnetic Amplifier (Figure M-1)

Figure M-1 shows the graphic symbol for a transductor, saturable-core inductor, or saturable-core reactor, all terms used for the basic circuit of a magnetic amplifier. The winding with five scallops is the control winding, the one with three scallops is the power winding. The saturable properties symbol superimposed on them indicates that they are magnetically coupled by a saturable core.

Figure M–1
Magnetic Amplifier

In the absence of a control signal the core does not saturate. The flux induced in the core rises toward the saturation level with each peak of the alternating current in the power winding, but does not reach it. The power winding, therefore, presents a continuous high impedance to the power supply, so that essentially all the power-supply potential is dropped across the power winding, and none across a load in series with it, as in Figure M-2 (a).

If a current is made to flow in the control winding, it will induce additional flux in the core. When the currents in the control winding and power winding are flowing in directions that cause the flux due to each winding to add, the total flux drives the core into saturation. When this happens the power winding becomes a very low impedance, and practically all the supply voltage appears across the load. This is illustrated in Figures M-2 (b) and (c).

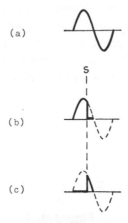

Figure M–2
AC Voltage Across Power Winding

The exact moment when the core saturates will depend upon the current level in the control winding. The flux induced by the power winding rises on each alternate half-cycle until saturation occurs, and this will be early or late in the half-cycle according to the level of flux induced by the control winding. Thus the power applied to the load can be set as desired by adjusting the level of the current in the control winding.

However, power can flow in the load only on every other half-cycle, since the core cannot saturate when the fluxes subtract. Consequently, full-wave power utilization requires that *two* saturable-core inductors be provided with their power windings connected oppositely, as shown in Figure M-3.

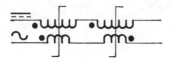

Figure M–3
Full-Wave Magnetic Amplifier

This also enables the saturated core to be reset to nonsaturation in time for the next half-cycle that saturates it. As the other core saturates, the current surge in its power winding induces a current pulse in its control winding. Since both control windings are connected in series, the pulse also appears in the other control winding, providing the necessary coercive force to reset the first core. See *Amplifier Circuits*.

Marker Generator (Figure M-4)

This is a simple method for generating markers on a frequency-response curve displayed on an oscilloscope screen. Crystals that are resonant at the desired marker frequencies are connected across the output of the sweep generator. Each crystal is an open circuit to all frequencies other than its own. When the sweep generator frequency coincides with its frequency, however, it shunts it to ground, resulting in a dip in the display. If the dip is too pronounced a suitable resistor should be connected in series with the crystal. See *Test and Indicating Circuits*.

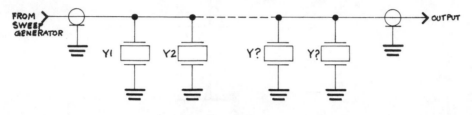

Figure M–4
Marker Generator

Matched Impedance Speaker Feed

See *Extension-Speaker Feed System*.

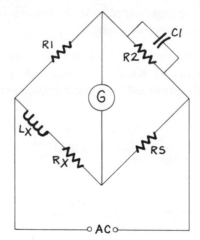

Figure M–5
Maxwell Bridge

Maxwell Bridge (Figure M-5)

The Maxwell bridge is an inductance-capacitance comparative device used for measuring unknown values of inductance. The value of L_x is found by:

$$L_x = R_1 R_s C_1$$

The value of the series resistance R_x is given by:

$$R_x = R_1 R_s / R_2$$

See *Test and Indicating Circuits*.

Megohmmeter (Figure M-6)

High values of resistance cannot be measured very well by an ohmmeter because of the limited reference voltage available. To obtain full-scale deflection when measuring hundreds of megohms requires thousands of volts. For this reason, special instruments called *megohmmeters*, or ''meggers,'' are used for determining the resistance of insulation, earth grounds, or other ultra-high resistances.

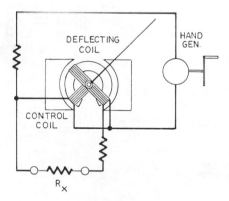

Figure M–6
Megohmmeter

Standard megohmmeters are capable of generating up to a thousand volts, and can indicate resistance values up to 2 gigohms. To generate the voltage, a hand generator is built into the instrument. The meter has a special type of crossed-coil movement, as shown in Figure M-6. The *deflecting coil* is in series with the generator and the unknown resistance. The *control coil* is in series with the generator only. Their inductances are equal.

With the input terminals shorted together, the current through each coil would be the same, but as their fields oppose each other the pointer does not move. When a resistance is connected across the input, the current through the deflecting coil changes, and the pointer now moves in proportion to the difference between the currents. The advantage of this arrangement is that the actual voltage of the generator output does not matter, since the meter is affected only by the change caused by the resistance.

In some megohmmeters the hand generator may be replaced with a regulated power supply capable of an output of many thousands of volts. The principle is still the same, however. See *Test Indicating Circuits*.

Mixer (Figure M-7)

This circuit is for an IGFET mixer in an FM tuner. The preceding RF amplifier stage and the local oscillator are not shown. The latter is tuned to be always 10.7 megahertz above the frequency of the incoming RF signal. Both signals are applied to the gate of the IGFET, and they appear in its drain circuit, together with the sum and difference frequencies, as described under *Converter*. It should be noted, however, that, unlike a bipolar transistor, there is no loading of the input signal when an IGFET is used. See *Demodulation Circuits*.

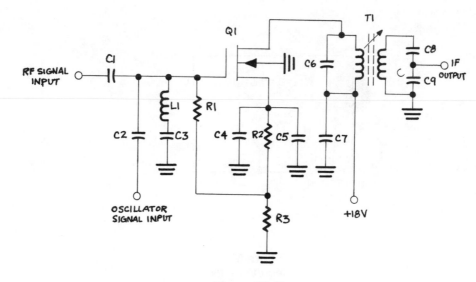

Figure M–7
Mixer

Parts List

C1	2.7 picofarads
C2	1.5 picofarads
C3	270 picofarads
C4	0.001 microfarad
C5	0.01 microfarad
C6	68 picofarads
C7	0.01 microfarad
C8	51 picofarads
C9	1200 picofarads

L1   1 microhenry (RF choke)
Q1   RCA–40559 or equivalent
R1   100 kilohms
R2   330 ohms
R3   680 ohms
T1   Double-tuned IF input transformer

MODULATION CIRCUITS

The transmission of information by radio requires the modulation of an electromagnetic wave in space. This *carrier wave* is a simple form of alternating current, practically a sine wave. A sine wave has three dimensions that can be modulated: its amplitude, its frequency, and its phase.

Amplitude modulation. In standard broadcasting and in the transmission of the picture portion of a television signal, amplitude modulation (AM) is used. As in telephony, the amplitude or strength of the carrier is varied in accordance with the amplitude and frequency of the modulating signal. For example, in transmitting a tone with a frequency of 400 hertz, the *amplitude* of the carrier is varied at a frequency of 400 hertz. The *level* of the amplitude variations depends upon the loudness of the tone.

Frequency modulation. In frequency modulation (FM) the amplitude of the carrier remains constant. The information to be transmitted is made to vary the carrier's *frequency* in accordance with the pitch and amplitude of the sounds to be transmitted. For example, in transmitting the same musical tone of 400 hertz the frequency of the carrier is made to vary above and below its normal frequency at a rate of 400 hertz. The amount by which the frequency varies is in proportion to the loudness of the tone.

Phase modulation. In phase modulation (PM) the phase of the carrier is advanced or retarded by variations in the amplitude of the information to be transmitted, and at a *rate* according to their frequency. The transmitter power does not vary but its phase does.

FM and PM are two forms of *angle modulation,* because the phase angle of the overall resultant signal swings back and forth with respect to the phase of the unmodulated carrier.

Sidebands. Modulation sets up new groups of radio frequencies above and below the carrier frequency, which are called sidebands. Consequently a modulated signal occupies a *band* of frequencies rather than the single frequency of the carrier alone. The carrier and its sidebands are called a *channel.* AM broadcasting stations are limited to modulated frequencies that must not exceed 5000 hertz. If a carrier of 1000 kilohertz is modulated with a 5000-hertz signal, an upper side frequency of 1005 kilohertz and a lower side frequency of 995 kilohertz will be produced. The maximum width of the channel is therefore 10 kilohertz for an AM station. Since the modulating signal seldom consists of a single tone, but rather of a wide range of audio frequencies, there will be many more side frequencies than two in the 10-kilohertz channel bandwidth.

In AM the upper and lower sidebands carry the same information; therefore, one of them is superfluous. Either could be filtered out at the transmitter. The same applies to the carrier, which contains no information at all. AM transmissions can be divided into:

Double sideband, full carrier (A3), the standard broadcast practice, in which both sidebands and the carrier are transmitted. This requires no special circuitry in the receiver, so is more economical for widespread use, but the power wasted at the transmitter makes it the least efficient system.

Single sideband, suppressed carrier (A3J), on the other hand, is the most efficient, since the transmitter does not radiate superfluous power. The receiver, however, has to generate and insert the missing carrier so that demodulation (recovery of the information) can take place. Single-sideband (SSB) transmission of this type is in common use for all forms of point-to-point communication.

Vestigial sideband (A5C) is the method used by television stations. As in the case of AM radio broadcasting, both sidebands and carrier must be transmitted if special circuitry in the receiver is to be avoided, but the modulation bandwidth of the picture information is 4.5 megahertz. When allowance is made for the sound signal and guard bands, the channel width for the whole transmission would be almost double the 6 megahertz allocated if both sidebands were transmitted in full. The lower sideband is therefore rolled off so that at and beyond 1.25 megahertz below the carrier frequency it is at least 20 decibels down from the carrier level.

Other types of sideband transmission used with with AM carriers are *single sideband, reduced carrier (A3A), two independent sidebands, reduced carrier (A3B), single sideband, reduced carrier (A4A and A7A),* and *two independent sidebands (A9B).* Only the first two are used for telephony in the ordinary sense. The others are used for facsimile and multi-channel telegraphy, or combinations of telegraphy and telephony, etc.

In FM and PM, sidebands are set up as in AM, but they are more complex. Even a single tone gives a whole series of pairs of side frequencies that are harmonically related to the tone frequency. The number of extra sidebands depends on the *modulation index,* which is given by *carrier frequency deviation* divided by *modulating frequency.*

In the case of FM, the larger the modulation index the more effectively this type of transmission performs in combating noise and interference. Also, the larger the modulation index the greater the channel bandwidth, which is approximately twice the frequency deviation plus twice the modulating signal frequency. However, sideband or carrier suppression is not possible, because the energy that goes into the sidebands is taken from the carrier, whereas in AM it is supplied by the modulation circuit.

Methods of modulation. Generally speaking, amplitude modulation consists of modulating the power supply or bias to some stage that affects the amplitude of the carrier. This is often the output stage, but can be any stage, beginning with the oscillator.

In AM, the degree to which the carrier is modulated depends upon the power of the modulating signal and is expressed as a percentage. In 100-percent modulation the carrier is driven all the way to zero at its minimum amplitude, and to double its unmodulated amplitude at the maximum value, as shown in Figure M-8.

In FM or PM, some means is required of making the carrier frequency vary in accordance with the modulating signal. Since the frequency is determined by the tuned circuit in the oscillator, circuits are used that vary the tuning electronically. Some examples of these circuits are listed below.

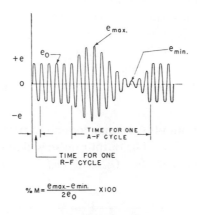

$$\% \, M = \frac{e_{max} - e_{min}}{2e_0} \times 100$$

Figure M–8
Modulation Percentage

Specific Modulation Circuits

See the following list for specific modulation circuits described elsewhere in the encyclopedia:

Absorption Modulator
Amplitude Modulation
Balanced Modulator
Cowan Bridge Balanced Modulator
Emphasis
Frequency-Modulated Modulator
Grid Modulation
Plate Modulator
Reactance Modulator
Ring Modulator
Solid-State Balanced Modulator
Suppressed-Carrier (Double-Sideband) Modulator
Tunnel-Diode AM Transmitter
Tunnel-Diode FM Transmitter
Varactor Frequency Modulator

Monostable Multivibrator (Figure M-9)

The monostable multivibrator's action is like that of a pushbutton switch. Normally it has no output, but if a signal is applied to the input a signal appears at the output, and remains as long as the input signal is present. This circuit is also called a single-shot or one-shot multivibrator, but its best-known form is the Schmitt trigger, shown in the figure.

Without an input signal, Q1 does not conduct, because its base is at zero potential, whereas its emitter is at a positive potential due to the current flowing in R5. On the other hand, Q2 is conducting because its base is more positive than its emitter. When an input signal with a voltage that overcomes the reverse bias of Q1 arrives on its base, Q1 turns on. This results in its collector potential falling almost to that of Q2's emitter, so that Q2's base is now reverse biased because of the voltage drops across R2 and R3. Q2 turns off, and its collector potential rises to the level of V_{CC}.

As long as the voltage of the input pulse is high enough to trigger Q1 it doesn't matter what its actual value is, or how noisy it is. The output pulse will always have the same potential, and noise will have been eliminated. See *Oscillator and Signal Generator Circuits.*

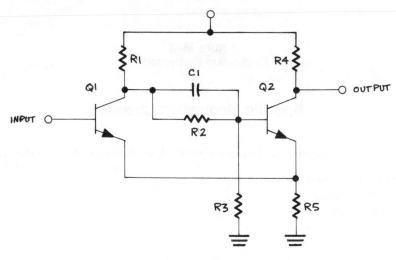

Figure M–9
Monostable Multivibrator

Parts List

C1	330 picofarads
Q1, Q2	2N706 or equivalent
R1	3.3 kilohms
R2	1.8 kilohms
R3	6.8 kilohms
R4	2.2 kilohms
R5	5.6 kilohms

Multimeter (Figure M-10)

A multimeter is an instrument that uses one meter or digital readout to measure several functions, such as DC voltage, AC voltage, current and resistance, selecting for each function and range by means of a multiposition switch. The principle is illustrated in the figure, although it should be noted that many small digital multimeters (DMMs) use a number of integrated circuits as well.

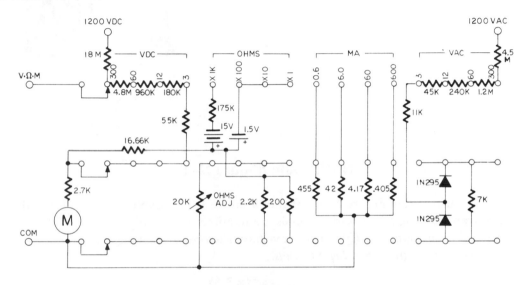

Figure M–10
Multimeter

The heart of this circuit is a sensitive d'Arsonval meter with a 50-microampere permanent-magnet moving-coil (PMMC) movement. A 1-microammeter meter would be more sensitive, and have less loading effect on the circuit in which a voltage measurement is made, but is too fragile for use in a rugged portable instrument. This circuit is a compromise, in which the maximum loading effect is 20,000 ohms per volt. Instruments are available that have a much higher input inpedance, obtained by using a field-effect transistor (FET) in the circuit.

Multimeters with a meter are often called VOMs (volt-ohm-milliammeter), with the term DMM being reserved for those with a digital readout. Parts are not listed, as the values are given in the diagram. See *Test and Indicating Circuits.*

Multivibrator

See *Astable Multivibrator, Bistable Multivibrator, Monostable Multivibrator,* or *Flip-Flop.*

NAND Gate (Figure N–1)

The circuit illustrated in Figure N–1 is that of a simple RTL NAND gate (RTL stands for "resistor-transistor logic"). Positive signals are required at both inputs before Q1 and Q2 can turn on. When this occurs, Q1's collector voltage falls to zero, resulting in an output inverted with respect to the inputs. See *Logic Circuits*.

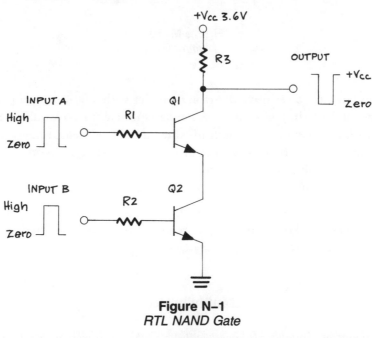

Figure N–1
RTL NAND Gate

Parts List

Q1, Q2	GE20 or equivalent
R1, R2	470 ohms
R3	640 ohms

For comparison, a CMOS NAND gate used in an integrated circuit is shown in Figure N-2. Q1 and Q2 are insulated-gate field-effect transistors (IGFETs) with P-channels (enhancement type). Q3 and Q4 are IGFETs with N-channels (enhancement type). [*Note:* IGFETs are very often called MOSFETs (metal-oxide-semiconductor FET).]

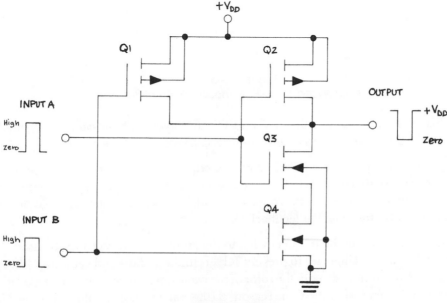

Figure N–2
CMOS NAND Gate

Negative Grid Bias

See *Bias Circuits.*

Neon Indicator

See *Pilot Lights.*

Neutralized Push-Pull Triode RF Amplifier
(Figure N–3)

Because of positive feedback from plate to grid, this circuit will function as an oscillator unless steps are taken to reduce or nullify the effect of the plate-grid capacitance in the two triode

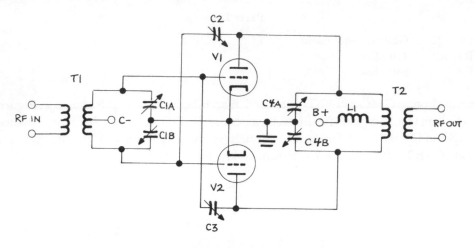

Figure N–3
Neutralized Push-Pull Triode RF Amplifier

tubes. This is done by connecting C2 and C3 as shown. These capacitors should have approximately the same value as the plate-grid capacitance of the triodes.

The midpoint of the plate tank circuit is grounded, so that the voltages at opposite ends of the tank are essentially equal, but 180 degrees out of phase. By cross-connecting them via the neutralizing capacitors to the opposite grids, as shown, the internal feedback is counterbalanced by the external feedback.

Parts values depend upon the frequency fed to the amplifier and type of tube used. L1 is an RF choke of 1 to 2.5 millihenries to prevent RF grounding of the coil centertap (through the power supply) simultaneously with the rotor of the capacitor, which would set up an additional unwanted tuned circuit. (For the same reason, a 100-ohm resistor should be installed in series with the C– supply.) See *Amplifier Circuits.*

Neutralized Transistor Amplifier (Figure N–4)

In high-frequency circuits, such as the IF amplifier shown in the figure, oscillation may occur due to positive feedback via the internal capacitance of the transistor. A neutralizing capacitor C3 may therefore be used (as in the triode amplifier described below) to feed back an equal signal of opposite phase to the transistor's base. See *Amplifier Circuits.*

Parts List

C1	0.05 microfarad
C2	9 picofarads
C3	0.05 microfarad
C4	0.05 microfarad

C5   220 picofarads
C6   0.02 microfarad
D1   1N34 or equivalent
Q1   GE82 or equivalent
R1   330 ohms
R2   560 ohms
R3   5.6 kilohms
R4   5 kilohms, variable
T1   IF transformer

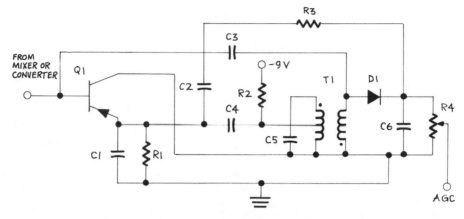

Figure N–4
Neutralized Transistor Amplifier

Neutralized Triode RF Amplifier (Figure N–5)

This circuit is basically the same as the transistor circuit above, except that it employs a triode tube. Negative feedback from T2's secondary applies a signal to V1's grid. The amplitude of the feedback is adjusted by C1 to counterbalance the plate-grid capacitance of the triode.

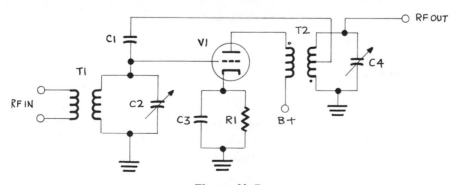

Figure N–5
Neutralized Triode RF Amplifier

C1 is adjusted by disabling V1 by removing the plate voltage. A signal is applied to its grid, and the signal level at the output is then measured with an RF voltmeter. C1 is adjusted for minimum output.

Parts values depend upon the frequency fed to the circuit and the type of triode employed. See *Amplifier Circuits*.

NOR Gate (Figure N–6)

The circuit illustrated in Figure N-6 is that of a simple RTL NOR gate (RTL stands for "resistor-transistor logic"). A positive signal is required at one of the inputs before Q1 or Q2 can turn on. When this occurs, all of the supply voltage is dropped across R2, resulting in an output inverted with respect to the input. See *Logic Circuits*.

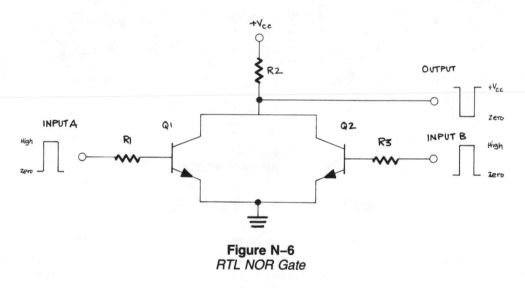

Figure N–6
RTL NOR Gate

Parts List

Q1, Q2	GE20 or equivalent
R1, R3	470 ohms
R2	640 ohms

For comparison, a CMOS NOR gate used in an integrated circuit is shown in Figure N-7. Q1 and Q2 are insulated-gate field-effect transistors (IGFETs) with P-channels (enhancement type). Q3 and Q4 are IGFETs with N-channels (enhancement type). [*Note:* IGFETs are often called MOSFETs (metal-oxide-semiconductor FET).]

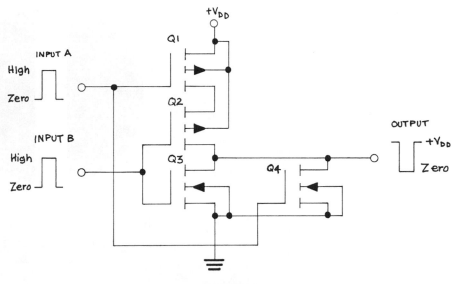

Figure N–7
CMOS NOR Gate

Notch Active Filter

See *Band Reject Filter*.

Notch Passive Filter

See *Passive Filter Circuits*.

One-Shot Multivibrator.

See *Monostable Multivibrator*.

OPERATIONAL AMPLIFIER CIRCUITS

An operational amplifier is a very high-gain, direct-coupled amplifier that uses feedback for control of its response characteristics. It usually consists of two cascaded differential amplifiers (q.v.), together with an appropriate output stage, as shown in Figure O-1. This is a stable configuration that lends itself well to the monolithic diffusion process used in microelectronics fabrication, and also provides high common-mode rejection at least 20 decibels greater than the differential gain. There are two input terminals: the inverting input, abbreviated II (or –), and the noninverting input, abbreviated NII (or +). The input signal and feedback are usually (but not invariably) connected to the inverting input, so that the output signal is of opposite phase to the input signal, and the feedback is negative.

This basic arrangement is shown in Figure O-2. The load resistor R_L is of a high enough value so that its effect on the transfer characteristics can be neglected (in other words, $I_O = 0$). In this case, the gain is given by $R_f/(R_i + R_s)$. To avoid any DC offset, DC paths for each input must be equal, so R_2 is made equal to $(R_i + R_s)$.

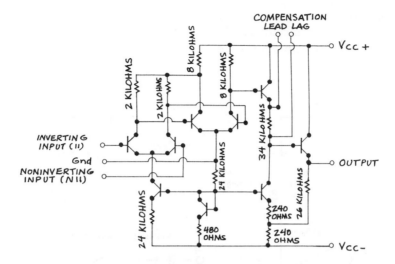

Figure O–1
Operational Amplifier

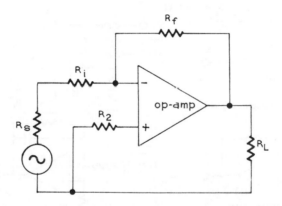

Figure O–2
Op-Amp Resistive Feedback Circuit

While a simple resistive feedback circuit has negligible phase shift at low frequencies, this is not the case at higher frequencies, so manufacturers generally provide phase-compensation terminals to which external components may be connected that modify the performance of the amplifier. This is called internal phase compensation. In one method, called straight roll-off, a suitable RC network is connected across an internal resistor. In another, called Miller-effect roll-off, the phase-compensating network is connected between the input and output of an inverting-gain stage. Many interesting circuits are given in the manufacturer's applications notes for various op-amps.

Several more op-amp circuits are given in Figure O-3. These are simplified by the omission of R_s, which is assumed to be included in R_i. R_L is also not shown, but its value must always be such that I_O is virtually zero. Op-amp performance may be calculated from the formulas given with each circuit.

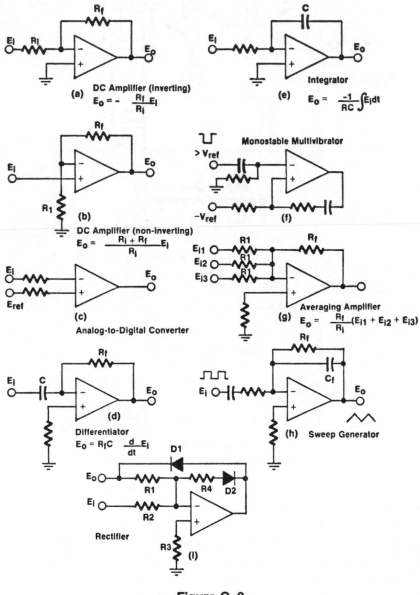

Figure O–3
Other Op-Amp Circuits

Other Operational-Amplifier Circuits

See the following list for other circuits using operational amplifiers described elsewhere in the encyclopedia:

ACTIVE FILTERS

Band-pass Active Filter
Band-Reject Active Filter
Bandstop Active Filter
High-Pass Active Filter

INTEGRATED CIRCUITS

Laboratory Power Supply
Phase-Shift Active Filter
Preamplifier with Operational Amplifier
Scaling Inverter with Operational Amplifier

OR Gate (Figure O-4)

This is an RTL OR gate. (RTL stands for "resistor-transistor logic.") When a positive input signal appears at Input A, Q1 turns on, so that the V_{CC} supply voltage is connected directly to the upper end of R2, resulting in a positive output. The same happens if a positive input signal is applied to Input B: Q2 turns on, connecting V_{CC} directly to the upper end of R2, so that a positive output results also. In short, a positive output is obtained whenever a positive input is applied to either input or both inputs. See *Logic Circuits.*

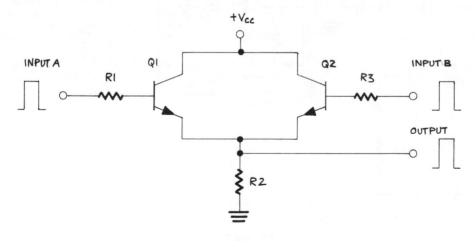

Figure O–4
RTL OR Gate

Parts List

Q1, Q2	GE20 or equivalent
R1, R3	470 ohms
2R	640 ohms

Oscillator and Frequency Doubler
(Figure O-5)

In this circuit, Q1 (a junction field-effect transistor or JFET) forms an oscillating circuit with the 14-megahertz crystal Y1. The output signal is coupled through T1 to the push-push circuit comprising Q2 and Q3. This amplifier automatically generates the second harmonic of the input signal frequency, to which T2 is tuned. See *Frequency Conversion Circuits; Oscillator and Signal Generator Circuits.*

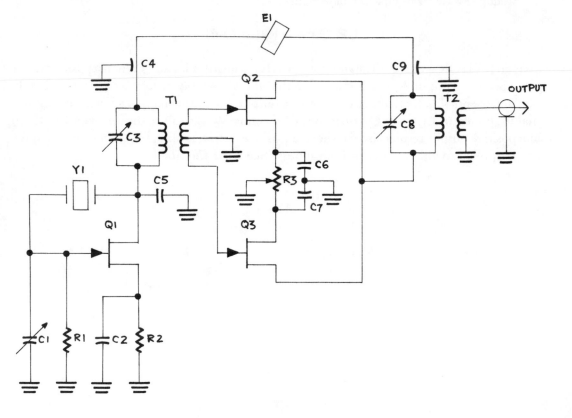

Figure O–5
Oscillator and Frequency Doubler

Parts List

C1	Variable, 10–60 picofarads
C2, C6, C7	0.01 microfarad
C3	Variable, 3–12 picofarads
C4, C9	1000 picofarads
C5	100 picofarads
C8	Variable, 8–35 picofarads
E1	Ferrite bead
Q1, Q2, Q3	MPF102, HEP802, or equivalent
R1	1 megohm
R2	1 kilohm
R3	2 kilohms, potentiometer ("balance control")
T1	RF coupling transformer, centertapped
T2	RF stepdown coupling transformer
Y1	14-megahertz crystal

OSCILLATOR AND SIGNAL GENERATOR CIRCUITS

If there is enough positive feedback in an amplifier circuit, self-sustaining oscillations will be set up. This is undesirable in an amplifier, but when an amplifier is arranged so that this condition exists, it is an oscillator.

There are two types of oscillator, *LC oscillators* and *RC oscillators*. In both types, a tube or transistor acting as an automatic switch chops the DC supply into a series of pulses, which are then shaped by other circuit elements. In LC oscillators the resonant circuit transforms the pulses into sine waves.

LC Oscillators. Oscillation normally takes place at only one frequency, and a desired frequency is obtained by using a resonant circuit tuned to that frequency. The frequency is given by the reciprocal of 6.28 times the square root of the product of the inductance (in henries) multiplied by the capacitance (in farads) ($1/2 \pi LC$). The proper phase for positive feedback can be obtained quite easily from a single tuned circuit. For example, in the Hartley oscillator in Figure O-6 the circuit C1-L1-L2 is tuned to the desired frequency. The coil L1-L2 is tapped, and the emitter of Q1 is connected to the tap. The base and collector are connected to opposite ends of the tuned circuit, so that L2, between the emitter and collector, is the output inductor; while L1, between the emitter and the base, is the input inductor. Signal energy that develops across L2 is induced into L1, so that coupling exists between the input and output circuits. As the two circuits are in phase, the conditions for oscillation exist.

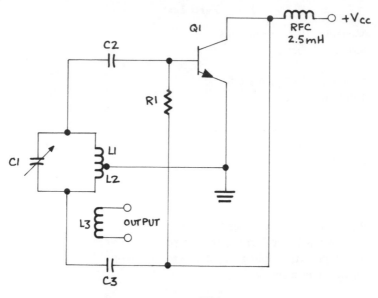

Figure O–6
Hartley Oscillator Feedback Circuit

When the upper end of L1 is positive with respect to the emitter, it charges C2 so that electrons are attracted from the base of Q1. These electrons cannot flow through L1 back to the emitter because C2 blocks direct current. They therefore have to flow or "leak" through R1, and in so doing cause a voltage drop across R1 that places a negative bias on the base. This cuts the transistor off so that current ceases to flow in L2. Since no current flows in L2, no current can be induced in L1, and the charge on C2 subsides. The base returns to the bias condition that allows Q1 to resume conduction, and the next cycle begins. This switching action takes place at the frequency of the resonant circuit. The current builds up and subsides *sinusoidally* because of the inductive and capacitive reactance of the resonant circuit.

The position of the centertap on L1-L2 is generally somewhere near the middle of the coil. If it is too near the base end, the voltage drop across L1 will be too small to give enough feedback; if it is too close to the collector end, the impedance between the emitter and collector will be too small to permit good amplification. Practically all feedback oscillators work in a similar manner, although their circuits may differ in details and appearance. The essential feature they all have is positive feedback in the proper amplitude to sustain oscillation.

RC Oscillators. The frequency of an RC oscillator is determined by the rate at which a capacitor can be charged and discharged by current through a resistance. The frequency is the reciprocal of the time constant, obtained by multiplying the capacitance (in farads) by the resistance (in ohms). Such an oscillator is also called a "relaxation oscillator," of which the simplest form is that shown in Figure O-7. The supply voltage charges C1 via R1 until the potential on C1 reaches the point where DS1 conducts. This causes the rapid discharge of C1, until its potential drops to the point where DS1 ceases conduction. C1's potential then builds up again for the next cycle.

This oscillator produces a sawtooth waveform, but it is not linear enough to be used in sweep generators and the like. For a linear sawtooth, circuits such as the phantastron or bootstrap configuration must be used.

A signal generator is a test instrument that contains an oscillator and is designed for generating signals for test purposes. The term is used somewhat loosely, since there are also audio oscillators and test oscillators for the same purpose. As a result, the term *signal generator* is more often applied to instruments that produce high-frequency signals, or functions other than sine waves (e.g., square waves, pulses, triangle waves, etc.).

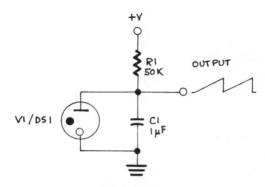

Figure O–7
Simple Relaxation Oscillator

Other Oscillator and Signal Generator Circuits

See the following list for other oscillator and signal generator circuits described elsewhere in the encyclopedia:

Armstrong Oscillator
Astable Multivibrator
Bistable Multivibrator
Blocking Oscillator
Bootstrap Sawtooth Generator
Bramco Resonator
Calibration Oscillator
Code-Practice Oscillator
Colpitts Oscillator, Crystal-Controlled
Colpitts Oscillator, Series-Fed
Colpitts Oscillator, Shunt-Fed

Electron-Coupled Oscillator
FET Oscillator and Frequency Doubler
Flip-Flop
Franklin Oscillator
Free-Running PRR Generator
Hartley Crystal-Controlled Oscillator
Hartley Oscillator
IC Bridged-T Oscillator
IC Function Generator
IC Phase-Shift Oscillator
Monostable Multivibrator
Oscillator and Frequency Doubler
Parallel-T Oscillator
Phantastron
Phase-Shift Oscillator
Pierce Crystal Oscillator
Pierce-Miller Crystal Oscillator
Quadrature Oscillator
Radio-Telegraph Transmitter
RC Oscillator
Relaxation Oscillator
Remote-Control Transmitter
Saturated Flip-Flop
Sawtooth Generator
Schmitt Trigger
Series-Tuned Emitter, Tuned-Base Oscillator
Transistor Audio Oscillator
Transistor Overtone Crystal-Controlled Oscillator
Transistor Phase-Shift Oscillator
Tunnel-Diode Oscillator
Tunnel-Diode Voltage-Controlled Oscillator
UJT Relaxation Oscillator
VHF Crystal-Controlled Oscillator
Wien Bridge Oscillator

Owen Bridge (Figure O–8)

The Owen bridge is used to determine the inductive and resistive components of an inductance. In the circuit shown, L_x is the inductive component of the unknown inductance, and its resistive component is denoted by R_x. To balance the bridge, C1 is adjusted for a null reading on the galvanometer. When the bridge is balanced, the unknown inductance is given by R1R2C2, and the unknown resistance is given by R2C2/C1. See *Test and Indicating Circuits*.

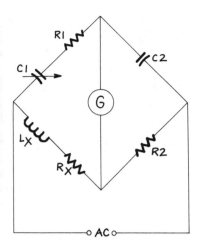

Figure O–8
Owen Bridge

Parallel Resonant Wavetraps (Figure P-1)

When an unwanted signal causes objectional interference to a TV or FM receiver with a balanced input, parallel wavetraps resonant at the frequency of the interfering signal should be connected as shown in the diagram. Parts values will depend upon the frequency of the unwanted signal. When tuned to that frequency, the two resonant wavetraps present an extremely high impedance to it, while allowing other signals to pass without noticeable attenuation. See *Antenna Circuits; Passive Filter Circuits.*

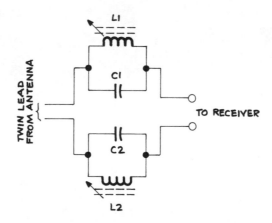

Figure P-1
Parallel Resonant Wavetraps

226

Parallel–T Oscillator (Figure P–2)

Part of the output of the emitter-follower Q2 is fed back to the base of Q1 via the parallel-T network C2, C4, R4, R5. The frequency of oscillation is varied by adjusting R7. See *Oscillator and Signal Generator Circuits.*

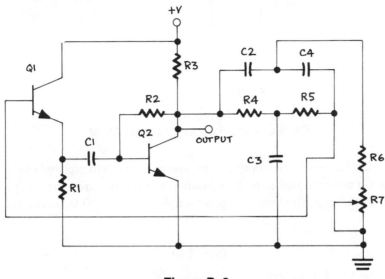

Figure P–2
Parallel-T Oscillator

Parts List

C1	0.1 microfarad
C2, C4	0.022 microfarad
C3	0.047 microfarad
Q1, Q2	GE20 or equivalent
R1, R6	4.7 kilohms
R2	180 kilohms
R3	3.3 kilohms
R4, R5	47 kilohms
R7	5 kilohms, variable (''frequency control'')

Passive Audio-Frequency Signal Mixer (Figure P–3)

This is a simple, inexpensive, but workable circuit for mixing two audio channels. The resistance values of R1 and R2 must equal the impedances of their signal sources, while the combined values of R3 and R4 in parallel should equal the input impedance of the load. It is possible,

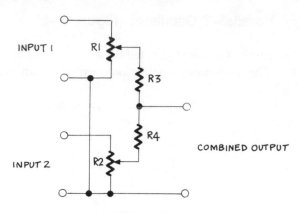

Figure P–3
Passive Audio-Frequency Signal Mixer

therefore, to connect microphones, phonographs, and so on, both to mix and match impedances. In the parts list it is assumed that two 50,000-ohm microphones are connected to Input 1 and Input 2, and that the output is connected to an amplifier with a 50,000-ohm impedance. See *Amplifier Circuits.*

Parts List

R1, R2   50 kilohms, variable
R3, R4   100 kilohms

PASSIVE FILTER CIRCUITS

The four basic filters are the low-pass, high-pass, band-pass, and bandstop filters. Although many refinements are possible, the four constant-K circuits given here are practical for most purposes and have the benefit of not requiring advanced mathematics to design them.

Low-Pass and High-Pass Filters. The low-pass filter is shown in Figure P-4, and its response curve in Figure P-5. The high-pass filter and its response curve are shown in Figures P-6 and P-7.

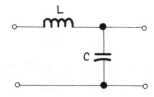

Figure P–4
Low-Pass Passive Filter Circuit

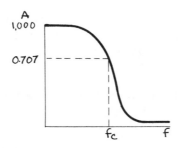

Figure P–5
Low-Pass Passive Filter Response Curve

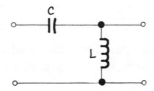

Figure P–6
High-Pass Passive Filter Circuit

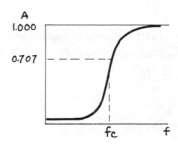

Figure P–7
High-Pass Passive Filter Response Curve

A low-pass filter acts as an AC voltage divider. The reactance of the capacitor must be such that it grounds all frequencies above the cutoff frequency. The inductor, on the other hand, offers a low reactance to frequencies below the cutoff frequency. This results in a high voltage drop across the capacitor and a low voltage drop across the inductor for those frequencies that are to be passed.

The high-pass filter works the other way round. The capacitor offers a low reactance to the higher frequencies, and the inductor offers a high reactance to them. This results in a low voltage drop across the capacitor and a high voltage drop across the inductor for those frequencies that are to be passed.

Values of C and L are calculated from the following formulas:

$$C = 1/6.28f_c R$$

$$L = R/6.28f_c$$

where:

      C   = capacitance in farads
      L   = inductance in henries
      R   = nominal terminating resistance, also equal to the square root of L/C
      f_c = cutoff frequency

RC Filters. In the low-pass and high-pass filters described above the inductor can be replaced by a resistor. The value of the resistor should equal that of the reactance of the inductor it is replacing.

When a resistor is used, resistance is substituted for reactance. Resistance is not frequency selective, so only the reactive element (the capacitor) can select for or against the frequency. Furthermore, the DC resistance of the resistor may be ten times the DC resistance of the inductor it replaces, so that power is wasted unnecessarily. An RC filter, therefore, is suitable only for low-current applications.

Band-Pass Filter. The band-pass filter is shown in Figure P-8, and its response curve in Figure P-9. Values of C_1, C_2, L_1, and L_2 are calculated from the following formulas:

$$C_1 = 6.28(f_2 - f_1)/(6.28f_o)^2 R$$
$$C_2 = 1/6.28R(f_2 - f_1)$$
$$L_1 = R/6.28(f_2 - f_1)$$
$$L_2 = 6.28R(f_2 - f_1)/(6.28f_o)^2$$

where:

      C_1 = series capacitance in farads
      C_2 = shunt capacitance in farads
      L_1 = series inductance in henries
      L_2 = shunt inductance in henries
      R = nominal terminating resistance, also equal to the square root of L_1/C_2, or the square root of L_2/C_1
      f_o = midband frequency
      f_1 = lower cutoff frequency
      f_2 = upper cutoff frequency

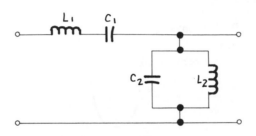

Figure P–8
Band-pass Passive Filter Circuit

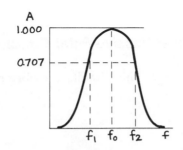

Figure P–9
Band-pass Passive Filter Response Curve

Bandstop Filter. The bandstop filter is shown in Figure P-10, and its response curve in Figure P-11. Values of C_1, C_2, L_1, and L_2 are calculated from the following formulas:

$$C_1 = 1/6.28R(f_2 - f_1)$$
$$C_2 = 248(f_2 - f_1)f_2f_1$$
$$L_1 = 6.28R(f_2 - f_1)/39.48f_2f_1$$
$$L_2 = R/6.28R(f_2 - f_1)$$

All expressions mean the same as for the high-pass filter.

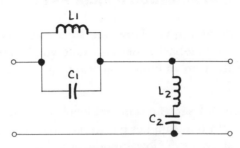

Figure P–10
Bandstop Passive Filter Circuit

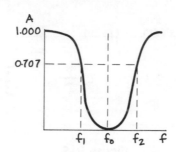

Figure P–11
Bandstop Passive Filter Response Curve

Other Passive Filter Circuits

See the following list for other passive filter circuits described elsewhere in the encyclopedia:

De-Emphasis
Differentiator
Double-Tuned Transformer
Dual Wavetrap
Emphasis
Parallel Resonant Wavetraps
Power Supply Circuits
Pre-emphasis
T-Network Filter
Video Equalizer

Pentagrid Converter

See *Converter*.

Phantastron (Figure P–12)

This circuit is a sawtooth generator (q.v.). However, since its principal use is in oscilloscopes, it is more usually known as a time-base generator. One good feature is that it produces its own gating waveform directly from an input triggering waveform, so that a separate circuit is not required.

When the circuit is quiescent, V1's screen conducts heavily, so that the screen voltage is low. The voltage divider R2-R3-R4 that supplies the screen voltage also supplies the suppressor voltage, so this is low as well. Since the low suppressor voltage prevents plate current from flowing, the plate voltage is high. This places a high voltage on the left-hand plate of C1.

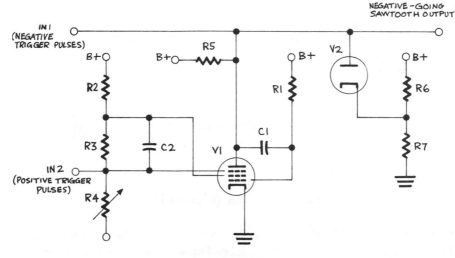

Figure P–12
Phantastron

When a negative-going trigger pulse at IN1 reaches V1's plate it is coupled through C1 to V1's grid. The grid is momentarily driven negative, cutting off the screen current. This causes the screen voltage to rise, and consequently the suppressor voltage also rises, since the increased potential reached it via R3. The increased suppressor voltage allows plate current to start, resulting in a drop in plate voltage. This lowers the potential on the left-hand plate of C1, and C1 discharges through R1, thus developing the rundown portion of a negative-going sawtooth or ramp.

Eventually the plate voltage drops so low that the plate no longer strongly attracts electrons, so the screen takes over. Now the screen voltage drops and the plate voltage rises. When it rises enough, V2 begins to conduct, clamping the plate voltage to a fixed quiescent DC level.

The suppressor supplies the positive gating waveform that starts and stops the plate current. This gating waveform is initiated by the trigger pulse, of course. If the trigger pulse is positive-going it can be applied directly to the suppressor via IN2. The rest of the action is as already described for the case when a negative-going pulse is applied to the grid. See *Oscillator and Signal Generator Circuits*.

Parts List

C1	Timing capacitor (value of C1 with that of R1 determines RC constant for each time base)
C2	0.001 microfarad
R1	(See above for C1)

R2, R6	680 kilohms
R3, R7	150 kilohms
R4	500 kilohms, variable ("stability control")
R5	270 kilohms
V1	6AU6 or equivalent
V2	Half of a 6AL5 or equivalent

Phase Discriminator

See *Discriminator*.

Phase Equalizer (Figure P–13)

This circuit is used to correct nonlinear phase shift in transmission lines and amplitude equalizers used in closed-circuit television. Parts values depend upon the application. However, the circuit is usually obtained as a commercial unit.

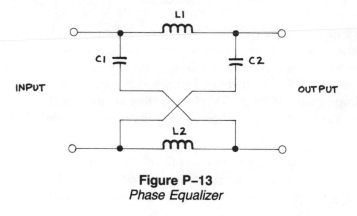

Figure P–13
Phase Equalizer

Phase Inverter (Figure P–14)

One type of phase inverter (also called a phase splitter) produces two output signals of opposite phase from a single-phase input signal, that can be used to drive a push-pull stage, for instance. Another type merely inverts the single-phase signal. (See *Inverter*.)

In a vacuum-tube circuit, a dual-triode tube, such as a 12AX7, may be used. The input signal is applied via C1 and R1 to one of the grids of the tube, and the output from that grid's plate is coupled via C3 to V2. An out-of-phase signal for the second grid of the 12AX7 tube is obtained from the voltage divider R5 and R6. The value of R6 is selected so that the signal fed back

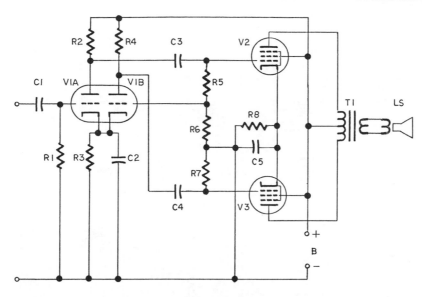

Figure P–14
Phase Inverter (Vacuum Tube)

from the junction of R5 and R6 will, when amplified by the second half of V1, have the same amplitude as the output signal from the first half. The signal amplified by the second half of the dual-triode is inverted with respect to the input, and is then applied via C4 to the grid of V3.

Parts List

C1	0.01 microfarad
C2	50 microfarads
C3, C4	0.047 microfarad
C5	20 microfarads
R1	500 kilohms, variable ("level control")
R2	3.9 kilohms
R3, R4	220 kilohms
R5, R7	470 kilohms
R6	7.5 kilohms (assuming gain of second half of 12AX7 is 60)
R8	82 kilohms
V1	12AX7 or equivalent
V2, V3	6BQ5 or similar output-amplifier tubes

In the transistor circuit of Figure P–15 an interstage coupling transformer with a split secondary is used to provide the required out-of-phase signal voltages for the push-pull inputs. See *Amplifier Circuits*.

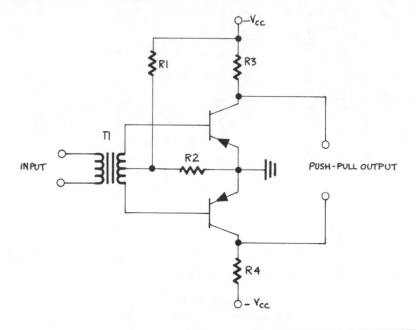

Figure P–15
Phase Inverter (Transistor)

Phase-Locked Loop Detector (Figure P–16)

A phase-locked loop would be very complex and expensive if fabricated of discrete components, but as it can now be obtained as an integrated circuit for negligible cost, it is used widely for modems, SCA and RTTY decoders, FM detection, frequency synthesis, and many other applications.

The figure illustrates in block form a typical IC (Signetics NE561B). An incoming FM signal is applied to a phase comparator, where it is compared with a signal generated by a voltage-controlled oscillator (VCO). The oscillator signal is at the resting frequency of the FM signal, so the frequency variations of the latter due to modulation result in error voltages that are amplified by the two amplifier stages, and become the audio output. A low-pass filter removes the RF component. Feedback to the VCO from the output of the first amplifier stage enables the VCO to phase-lock to the incoming signal to provide for more precise tuning.

An AM signal is fed to both the phase comparator and a frequency multiplier. The VCO locks on to this signal, and generates an unmodulated signal at the AM carrier frequency.

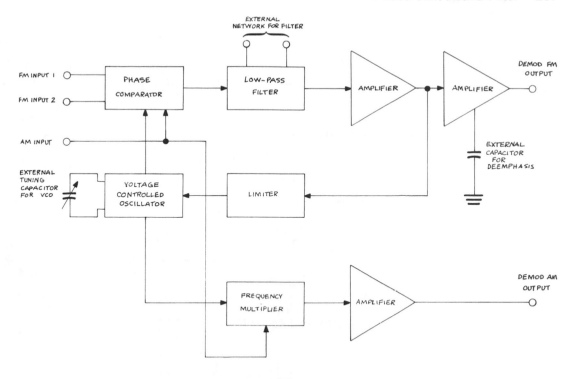

Figure P–16
Phase-Locked Loop Detector

This signal is then fed to the frequency multiplier, where it is compared with the modulated signal. The only difference between them is the varying modulation voltage, and this is separated from the RF (which is filtered out) and amplified to give the audio output. See *Demodulation*.

Phase-Shift Active Filter (Figure P–17)

A phase-shifting filter passes all frequencies with equal amplitude, but shifts their phase according to their frequency. It is therefore also called an all-pass filter. In this circuit, the value of C1 is found by dividing the frequency of the signal to be phase-shifted into 10, which gives C1's capacitance in microfarads. C2's capacitance is the same as C1's. This circuit will give a gain of 0.75 with the following resistor values for the phase-shift angles listed. See *Active Filter Circuits*.

Figure P–17
Phase-Shift Active Filter

PHASE SHIFT	R1	R2	R3	R4
10°	30.97	371.65	495.52	1486.57
20	16.31	195.84	261.13	783.36
30	11.69	140.30	187.06	561.19
40	9.49	113.99	151.97	455.93
50	8.24	98.88	131.84	395.51
60	7.43	89.12	118.82	356.48
70	6.85	82.28	109.71	329.13
80	6.44	77.19	102.94	308.79
90	6.11	73.24	97.64	292.91
−10	0.68	8.16	10.89	32.66
−20	1.29	15.49	20.65	61.96
−30	1.80	21.62	28.83	86.50
−40	2.21	26.62	35.49	106.47
−50	2.56	30.69	40.91	122.73
−60	2.83	34.04	45.39	136.17
−70	3.07	36.87	49.16	147.47
−80	3.28	39.30	52.39	157.19
−90	3.45	41.44	55.24	165.71

Phase-Shift Oscillator (Figure P–18)

For low frequencies, sine wave oscillators can be designed without inductors. In this circuit an RC network consisting of three L-type sections in tandem provides a 180-degree phase shift at 55 hertz. With three sections a current gain of 56 is sufficient to ensure oscillation. See *Oscillator and Signal Generator Circuits*.

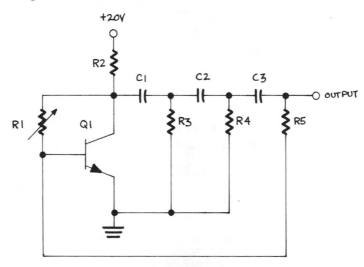

Figure P–18
Phase-Shift Oscillator

Parts List

C1, C2, C3	0.2 microfarad
Q1	2N3568 or equivalent
R1	500 kilohms, variable
R2	10 kilohms
R3, R4, R5	5 kilohms

Pierce Crystal Oscillator (Figure P–19)

In this oscillator there is no tuned circuit. The crystal is the equivalent of a high-Q tuned circuit of fixed frequency, and the oscillator will work with a wide range of crystal frequencies without change in the values of the other components. Feedback is by interelectrode capacitance. Capacitor C1 may not be required if grid-cathode capacitance is sufficient. See *Oscillator and Signal Generator Circuits*.

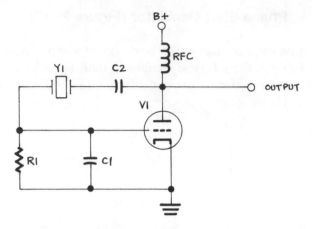

Figure P–19
Pierce Crystal Oscillator

Parts List

C1   50–100 picofarads (see text)
C2   0.001 microfarad
R1   50 kilohms
RFC   2.5 millihenries
V1   6C4 or equivalent
Y1   Crystal for desired frequency

Pilot Lights (Figure P–20)

A pilot light indicates whether a circuit is functioning or energized. It is a common practice to provide a pilot light that operates in conjunction with a power switch to indicate that the equipment is "on." This makes it a safety device as well.

Pilot lights may be small incandescent lamps, neon lamps, or light-emitting diodes (LEDs). Incandescent lamps can be used with lenses of any color, but neon lamps and LEDs are restricted in this respect because of their own color.

Due to their higher operating voltage, neon lamps are generally connected across the power line with a suitable dropping resistor, as shown in (A). Incandescent lamps and LEDs operate at much lower voltages, and so are usually connected across a secondary winding of a power transformer, as in (B). See *Test and Indicating Circuits.*

Pi-Network Coupler (Figure P–21)

When this coupler is used to connect the output of a transmitter to its transmission line, C1 is tuned to make the circuit resonant at the transmitting frequency, and C2 is tuned to match the network to the line. Values of the components will depend upon the frequencies and impedances involved. See *Amplifier Circuits; Antenna Circuits.*

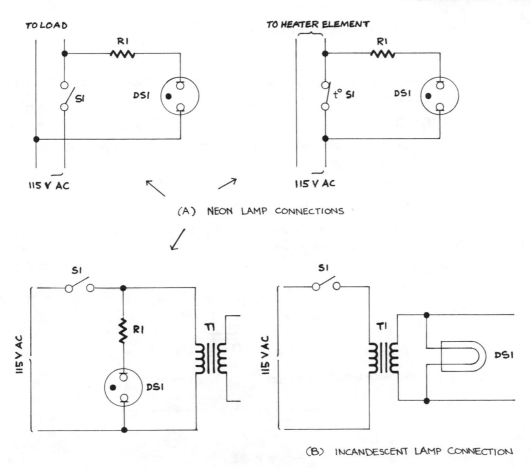

(A) NEON LAMP CONNECTIONS

(B) INCANDESCENT LAMP CONNECTION

Figure P–20
Pilot Lights

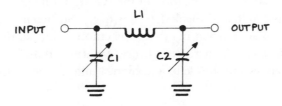

Figure P–21
Pi-Network Coupler

Pi-Section Filter (Figure P–22)

This is a popular type of capacitor-input filter used in power supplies as a ripple filter. The value of C1 is given as:

$$C1 = 0.00188/rR_L$$

where: r = maximum RMS voltage across C1 that can be accepted, divided by the DC voltage across C1.

R_L = maximum value of total load resistance in megohms.

The value of C1 should not be less than 4 microfarads.

The minimum value of L1 in henries is determined by dividing R_L in ohms by 1000. The value of C2 in microfarads is then found from:

$$C2 = 0.83/rL1$$

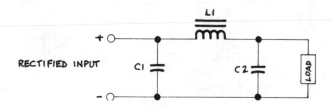

Figure P–22
Pi-Section Filter

Plate Detector (Figure P–23)

This circuit is arranged so that rectification of the RF signal takes place in the plate circuit of the tube, as contrasted to the grid-leak detector (q.v.). Sufficient negative bias is supplied to the grid to bring the plate current nearly to the cutoff point, so that the application of a signal to the grid circuit causes an increase in average plate current. The average plate current follows the changes in signal amplitude in a fashion similar to the rectified current in a diode detector.

The circuit illustrated is for a triode, but a circuit for a pentode is exactly the same, with the addition of a screen dropping resistor and bypass capacitor. See *Demodulation Circuits*.

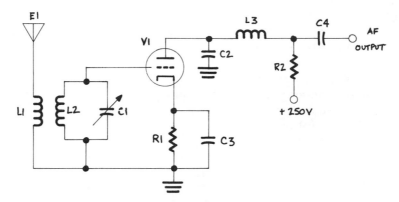

Figure P–23
Plate Detector

Parts List

C1, L1	Input circuit tuned to signal frequency
C2	0.5 microfarad or larger
C3	0.001 to 0.002 microfarad
C4	0.1 microfarad
R1	25 to 150 kilohms
R2	50 to 100 kilohms
RFC	2.5 millihenries
V1	Any suitable triode (e.g., 6C4)

Plate Modulator (Figure P–24)

This circuit illustrates the most widely used system of plate modulation. V2 and V3 form the output stage of an audio power amplifier. The AF power generated by this stage is combined with the DC power in the RF output stage (V1) through the coupling transformer T1. For 100 percent modulation the voltage on the plate of V1 must be able to vary between zero and twice the DC operating plate voltage, so V2 and V3 have to supply at least 50 percent of the power of the final RF stage.

Parts List

C1	47 picofarads, 1000 volts, mica
C2, C3, C5	0.01 microfarad
C4	Neutralizing capacitor (National STN or equivalent)
C7	Transmitting variable, 100 picofarads per section
L1, L2	2.5 millihenry RF chokes

L3, L4	Output tank, according to frequency band used (e.g., L3 would be 1.8 microhenries for 28 MHz)
R1	150 ohms
R2	400 ohms, 40 watts
T1	Output transformer, UTC VM-3, or equivalent
V1	5514 or equivalent
V2, V3	805 or equivalent

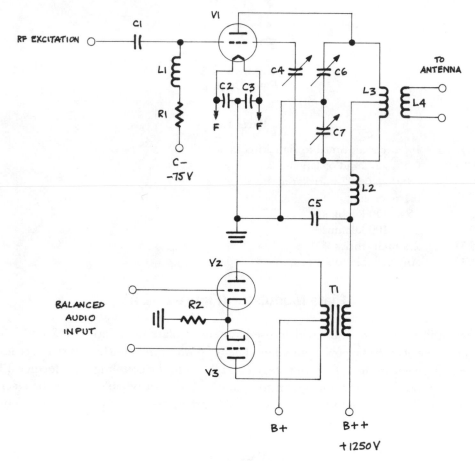

Figure P–24
Plate Modulator

Powerline Filter (Figure P–25)

Certain types of equipment are susceptible to noise, transients, and other interference on the power line. This is specially true of hi-fi systems and computers. To eliminate unwanted signals, low-pass filters are used that attenuate everything over 60 hertz.

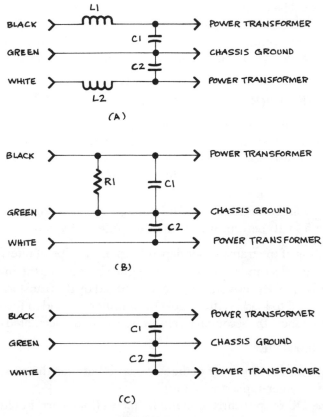

Figure P–25
Powerline Filter

The best type uses LC circuits in each powerline conductor, as shown in (A). Typical values would be 25 millihenries for the inductors and 0.5 microfarad for the capacitors. However, exact figures would take into account the source and load impedances.

A filter using two capacitors and a resistor, shown at (B), would require each capacitor to have the value 0.01 microfarad, and the resistor a value of 100 kilohms.

The simplest filter is shown at (C), with two capacitors of 0.1–0.5 microfarad.

POWER SUPPLY CIRCUITS

The power required by transistors and electron tubes is DC. For many purposes this is obtained from batteries, but for equipment with more than minimal power requirements it is the universal practice, where power lines are available, to obtain DC by means of a transformer-rectifier-filter system called a *power supply*. (There are also power supplies that manage without a transformer. These are called transformerless supplies and are described also in this article.)

Power Transformers. The output voltage that the power transformer must deliver depends upon the required DC load voltage and the type of rectifier circuit. With a capacitor-input filter, the RMS secondary voltage is usually made equal to or slightly more than the required DC output voltage. With a choke-input filter the required RMS secondary voltage can be calculated from:

$$E_t = 1.1[E_o + I(R_1 + R_2)/1000 + E_r]$$

where: E_o = the required DC output voltage
 I = the load current (including bleeder current, if any), in milliamperes
 R_1 = DC resistance of the first choke
 R_2 = DC resistance of second choke (if any)
 E_r = voltage drop in the rectifier
 E_t = the full-load RMS secondary voltage; the open-circuit voltage will usually be 5 to 10 percent higher than the full-load value

The volt-ampere rating of the transformer depends upon the type of filter. With a capacitor-input filter the heating effect in the secondary is higher because of the high ratio of peak to average current; consequently the volt-amperes consumed by the transformer may be several times the watts delivered to load. With a choke-input filter, provided the input choke has at least the critical inductance, the secondary volt-amperes can be calculated quite closely from:

$$VA_{sec} = 0.00075EI$$

where: E = the *total* RMS voltage of the secondary (between the outside ends in the case of a center-tapped winding)
 I = the DC output current in milliamperes (load current plus breeder current)

The primary volt-amperes will be 10 to 20 percent higher because of transformer losses.

Rectifier Circuits. Figure P-26 shows the half-wave rectifier circuit, which allows current to flow through the rectifier when it is forward biased, but not during the half of the AC cycle when it is reverse biased. The output wave, therefore, consists of positive-going pulses separated by intervals of the same width as the pulses. The output voltage with this circuit is 0.45 times the RMS value of the AC voltage delivered by the transformer secondary. Because the frequency of the pulses in the output wave is relatively low, considerable filtering is required to provide adequately smooth DC output. For this reason, this circuit is usually limited to applications where the current involved is small.

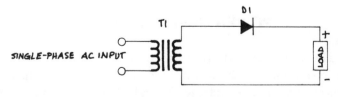

Figure P–26
Half-Wave Rectifier

Full-Wave Centertap Rectifier. This circuit (Figure P-27) is essentially a combination of two half-wave rectifiers. They conduct alternately, so that both halves of the AC cycle are used. When the upper end of the transformer's secondary is positive, D1's anode is positive also, so D1 is forward biased. This condition allows current to flow from the centertap of the secondary via the load and D1 to the upper end of the secondary. When the upper end of the secondary is positive the lower end is negative, so D2's anode is negative also. Consequently D2 is reverse biased and cannot conduct, so no current flows from the centertap and load through D2. When the polarity reverses on the next half-cycle, D2's anode becomes positive and D2 conducts; D1 is now reverse biased and cannot conduct. Current now flows through the load and D2 to the lower end of the secondary winding.

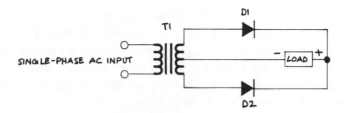

Figure P–27
Full-Wave Centertap Rectifier

The average output voltage is 0.9 times the RMS value of the voltage across *half* of the transformer secondary. For the same *total* secondary voltage, the average output voltage will be the same as that delivered by a half-wave rectifier. However, the frequency of the output pulses is twice that of the half-wave rectifier, so much less filtering is required. Since the rectifiers work alternately, each handles half of the average load current. The load current which may be drawn from this circuit is twice the rated load current of a single rectifier.

Full-Wave Bridge Rectifier. In this circuit (Figure P-28) two rectifiers operate in series on each half of the cycle. When the upper end of the secondary is positive, current flows from its negative lower end via D4, through the load, and back to the upper end via D1. When the lower end is positive and the upper end negative, current flows from the upper end via D3, through the load, and back to the lower end via D2. The output waveshape is the same as that from the centertap circuit. The output voltage obtainable with this circuit is 0.9 times the RMS voltage delivered by the transformer secondary. For the same total transformer secondary voltage, the average output voltage will be twice that obtainable with the centertap rectifier circuit. However, when comparing rectifier circuits for use *with the same transformer,* it should be remembered that the *power* which a given transformer will handle remains the same regardless of the rectifier circuit used. If the output voltage is doubled by substituting a bridge circuit for a centertap circuit, only half the rated load current can be taken from the transformer without exceeding its normal rating. Consequently the value of load current which may be drawn from the bridge rectifier circuit is still only twice the rated DC load current of a half-wave rectifier.

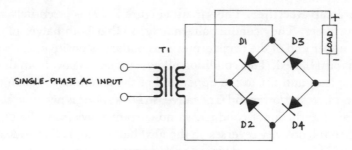

Figure P-28
Full-Wave Bridge Rectifier

Rectifiers. Semiconductors have replaced electron tubes in most rectifier applications. Silicon, germanium, selenium, and copper oxide are the semiconductors used. They have the advantage of not requiring filament power, and for a given power rating are smaller in size, even when mounted on heat sinks. However, electron tubes do retain one advantage: the reverse resistance of a tube rectifier is virtually infinite, whereas that of a semiconductor, while typically at least 10^3 times the forward resistance, is definitely finite.

Silicon rectifiers are used more than any other. They are small, light, and have a long life. However, they have no ability to recover from voltage transients (selenium is best in this connection), and their thermal capacity is poor (copper oxide is the winner here). The maximum operating temperature of a silicon diode is also the highest (200°C).

Filters. The DC pulsations produced by the rectifier circuits must be smoothed out, so a filter consisting of chokes and capacitors is connected between the rectifier output and the load circuit. The filter makes use of the energy-storage properties of the inductance of the choke and the capacitance of the capacitor, storing energy during the period in which the voltage and current are rising and releasing it to the load circuit while the amplitude of the pulse is falling, thus leveling off the output by both lopping off the peaks and filling in the valleys.

The pulsations in the output of the rectifier can be considered to be the resultant of an alternating current superimposed upon a steady direct current. The filter may then be thought of as a shunting capacitor that short-circuits the AC component while not interfering with the flow of the DC component, and a series choke that passes DC readily but which impedes the flow of the AC component.

The AC component is called the ripple. The effectiveness of the filter can be expressed in terms of percent ripple which is the ratio of the RMS value of the ripple to the DC value in terms of percentage. To avoid objectionable hum this percentage should not exceed 0.1.

The frequency of the pulsations in the rectifier output is called the ripple frequency. For a half-wave rectifier this is 60 hertz; for a full-wave rectifier it is 120 hertz. More filtering is required for 60 hertz than for 120 hertz. Figure P-29 shows a choke-input filter, so called because the first filter element is a choke. In a capacitor-input filter (Figure P-30) the first element is a capacitor.

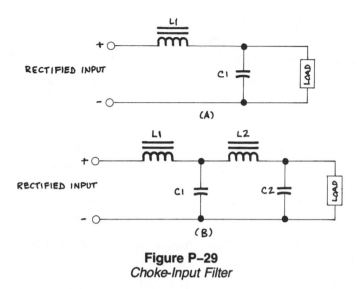

Figure P–29
Choke-Input Filter

In a single-section filter for a full-wave rectifier the percentage ripple r is given by

$$r = 100/LC$$

where L is in henries and C is in microfarads. In a half-wave rectifier the inductance and capacitance values must be doubled.

Voltage Regulation. Unless the power supply is designed to prevent it, there may be a considerable difference between the output voltage when it is unloaded and when a load is applied. Regulation is commonly expressed as the percentage change in output voltage:

$$\text{Regulation} = 100\,(E_1 - E_2)/E_2 \text{ percent}$$

Choke-input filters provide good voltage regulation when conservatively-rated components are used. This is more important when a variable load is involved. It is generally of less importance in receivers or other equipment where the load does not change.

Bleeder Resistor. In supplies where the output is over 50 volts a resistor may be connected across the output of the filter. Its main purpose is to discharge the filter capacitors when the supply is turned off. Its value in ohms should be 1000 times the maximum inductance of the choke in henries.

Input Choke. The value of the input-choke inductance which prevents the DC output voltage from rising above the average of the rectified AC voltage is called the *critical inductance* (L_{crit}). For a full-wave rectifier it is given by

$$L_{crit} = R_L/1000$$

where R_L is the load resistance in ohms. An inductance of twice the critical inductance is called the *optimum value,* and should be used for best results.

Swinging Chokes. Since the critical inductance varies with the load resistance, chokes called "swinging chokes" are used where much variation occurs. A swinging choke has the value of the critical inductance at minimum load and the optimum value at full load.

Filter Capacitor. The value of the filter capacitor can now be obtained by going back to the formula for percentage ripple and rearranging it:

$$C = 100/rL$$

Resonance. The values chosen for choke and capacitor should not be such that a circuit resonant at the ripple frequency is obtained, for then the effect would be the opposite of that for which the filter is designed. In the case of a full-wave rectifier this would happen when the product of choke inductance in henries and capacitor capacitance in microfarads is equal to 1.77. At least twice this product should be used to avoid resonance effects.

Output Voltage. Provided the input-choke inductance is at least the critical value, the output voltage is given by:

$$E_o = 0.9E_t - [(I_b + I_L)(R_1 + R_2)/1000] - E_r$$

where: E_o = the output voltage
  E_t = the RMS voltage applied to the rectifier (in the case of a centertapped secondary, this is the voltage between the centertap and one end of the secondary)
  I_b = the bleeder current in milliamperes
  I_L = the load current in milliamperes
  R_1 = the resistance of the first filter choke
  R_2 = the resistance of the second filter choke (if any)
  E_r = the potential drop across the rectifier

Ripple Reduction Factor. The ripple reduction factor is obtained from Figure P-31. For example, if L is 10 henries and C is 8 microfarads, then $L \times C = 80$, which gives a ripple reduction factor of 0.017.

Additional Filtering. If two filter sections are used, the total ripple reduction is equal to the product of the two individual ripple reduction factors, assuming both sections have the same configuration.

Capacitor-Input Filters. Compared to a properly designed choke-input filter, the DC output voltage of a capacitor-input filter (Figure P-30) is higher for most loads, the ratio of peak rectifier current to DC output current is greater, and voltage regulation is considerably poorer. Nevertheless, capacitor-input filters are often preferred because of their smaller size and weight.

The design of a capacitor-input filter starts with the degree of filtering required, which for a full-wave rectifier is given by:

$$r = (0.00188/C_1 R_L)$$

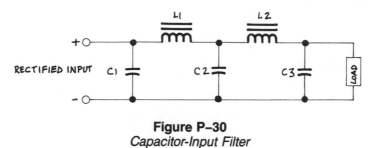

Figure P–30
Capacitor-Input Filter

where: C_1 = input-capacitor capacitance

R_L = load resistance

C_1R_L is in microfarads × megohms or farads × ohms. Multiply by 2 for a half-wave rectifier.

The input-capacitor capacitance must not exceed the value that gives the maximum allowable peak-current rating for the rectifier.

The approximate performance of the capacitor-input filter shown in Figure P-32 is best determined from the charts given in Figures P-31, P-33, P-34, and P-35. This is a pi-filter in which:

L1 = 20 henries
C1 = 4 microfarads
C2 = 8 microfarads
R_i = 200 ohms
E_r = 400 volts RMS
R_L = 5000 ohms

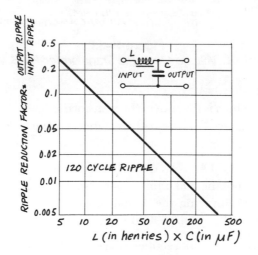

Figure P–31
Ripple Reduction Factor

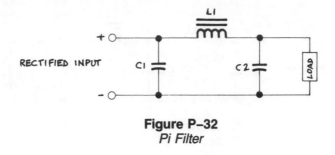

Figure P–32
Pi Filter

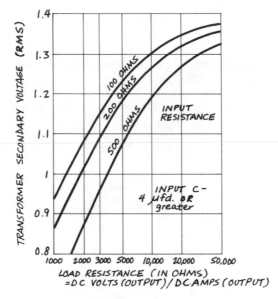

Figure P–33
Ratio of DC Voltage Output to Transformer RMS Voltage

R_i is the input resistance, which is the sum of the transformer and rectifier resistances. E_r is the transformer secondary voltage. R_L is the load resistance, and includes the resistance of the filter choke.

From Figure P-33 the ratio of DC volts output to transformer RMS volts—1.17—is given by the point of intersection of the 200 ohms input resistance curve with the 5000 ohms load resistance vertical axis. Multiplying the actual transformer RMS voltage E_r by this ratio gives $400 \times 1.17 = 468$ volts for the DC output.

From Figure P-34, the ratio of peak rectifier current to DC load current is similarly found to be 4. The DC load current $= 468/5000 = 93.6$ milliamperes, so the peak rectifier current is $93.6 \times 4 = 374$ milliamperes.

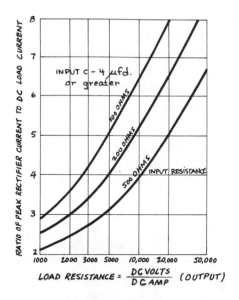

Figure P–34
Ratio of Peak Rectifier Current to DC Load Current

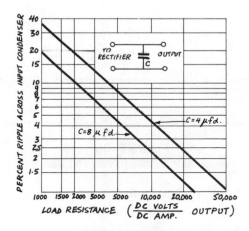

Figure P–35
Ripple Percentage

From Figure P-35, the ripple percentage across the input capacitor (4 microfarads) for a load resistance of 5000 ohms is approximately 8 percent.

L × C = 20 × 8 = 160, which in Figure P-31 gives a reduction factor of 0.009. The total ripple reduction, therefore, will be that across the input capacitor times the reduction factor for the second section (L1C2) = 8 × 0.009 = 0.072 percent.

Ratings of Filter Components. The minimum safe capacitor voltage rating should be not less than the peak transformer voltage. For instance, if the secondary of the transformer delivers 100 volts RMS, the minimum safe voltage rating is $100 \times 1.41 = 141$ volts. The practical rating should be somewhat higher for safety. Chokes should not only be able to handle the maximum DC load current, but their windings should be adequately insulated from the core to withstand the full DC output voltage of the supply, since the winding is in the positive lead and the core is grounded.

Regulation. In addition to the preceding circuits, regulation circuits are provided in many power supplies. These circuits ensure a constant output regardless of variations in load resistance or line voltage. The principle of these is shown in block diagram form in Figure P-36.

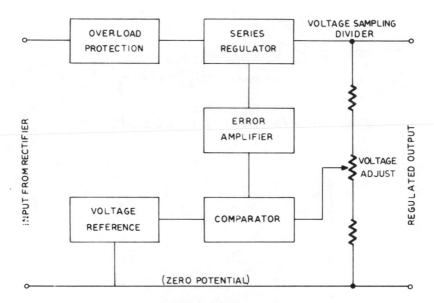

Figure P–36
Regulator Circuit (Block Diagram)

If the output voltage should rise or fall from the value set by the adjustment of the voltage-adjust potentiometer, the change is sensed by the comparator, which monitors the output voltage constantly against the voltage reference (frequently a zener diode). An error voltage is generated, amplified, and used to adjust the bias on the series regulator to restore the original level.

When a regulator circuit is used, the power supply filter circuit can be considerably simplified or dispensed with altogether, since DC pulsations sensed by the comparator are automatically ironed out by the error signal sent to the series regulator.

A basic series regulator is shown in Figure P-37. D1 is the reference. The output is sampled by the voltage divider R3-R4-R5. R4 is adjustable for varying the output or for calibration. The difference between the zener voltage and that picked off by the movable contact of R4

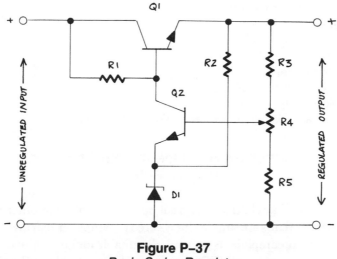

Figure P–37
Basic Series Regulator

determines the base-emitter voltage of Q2, which in turn controls Q2's conduction. Variations in the collector current of Q2 result in variations in the voltage drop across R1. These voltage variations are also present on the base of Q1, so that Q1's conduction rises or falls accordingly. As a result, the output voltage is held constant.

Other Power Supply Circuits

See the following list for other power supply circuits described elsewhere in the encyclopedia:

AC Adapter
Battery Eliminator
Bridge Rectifier
Bridge Voltage Doubler
Capacitor-Input Filter
Cascade Voltage Doubler
Choke-Input Filter
Chopper
Dual Voltage Power Supply
Full-Wave Rectifier
Half-Wave Rectifier
High-Voltage Power Supply
Instant-On Radio or TV Control
Laboratory Power Supply
Line-Voltage Adjuster
Pi-Section Filter
Powerline Filter
Resistor-Input Filter
Series Voltage Regulator with Comparator

Series Voltage Regulator with Preregulator
Shunt Voltage Regulator
Switching Regulator
Three-Phase Rectifier
Transmitter Power Supply
Universal AC-DC Power Supply
Voltage Doubler
Zener-Diode Voltage Regulator (see Series Voltage Regulator)

Preamplifier with Operational Amplifier
(Figure P-38)

This circuit uses half of a Fairchild μA739 dual op-amp. When the other half is used in an identical circuit, the two circuits are suitable for a stereo preamplifier that gives about 60 decibels channel separation. This preamplifier is particularly suitable for tape input or other application where a low-noise, high-gain preamplifier is required. See *Amplifier Circuits.*

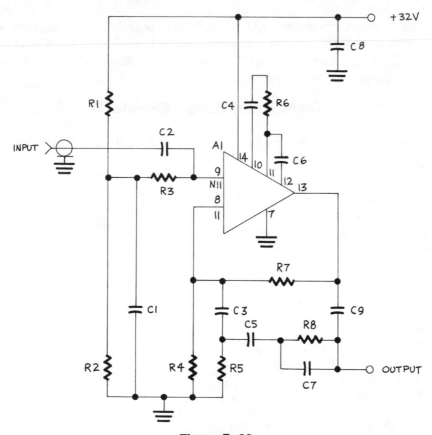

Figure P–38
Preamplifier with Operational Amplifier

Parts List

A1	1/2 of a Fairchild μA739 or equivalent
C1, C2, C3, C9	5 microfarads
C4, C6	0.005 microfarad
C5	0.0027 microfarad
C7	150 picofarads
C8	0.01 microfarad
R1	470 kilohms
R2	47 kilohms
R3	150 kilohms
R4	270 kilohms
R5	1 kilohm
R6	10 ohms
R7	1 megohm
R8	100 kilohms

Pre-Emphasis (Figure P-39)

This circuit is used in FM transmission to provide more amplification of the higher audio frequencies than of the lower audio frequencies. This is done because the higher frequencies do not have the amplitude of the lower. If they were transmitted without pre-emphasis the noise level would be higher in relation to the signal. In this filter circuit, the reactance of the inductor is greater at higher frequencies, so that a higher voltage is developed across it for higher frequencies.

The standard time constant for these circuits is 75 microseconds. In this example it is given by:

L (in henries)/R (in mcgohms) = 7.5/0.1 = 75 microseconds

At the receiver a de-emphasis circuit (q.v.) restores the normal balance between the frequencies. See *Passive Filters*.

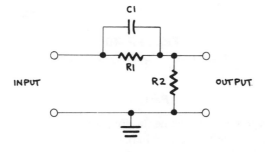

Figure P–39
Pre-Emphasis

Probe

See *High-Voltage Probe; RF Probe.*

Push-Pull Amplifier

See *Audio Power Output Stage; RF Amplifier with Two Triodes in Push-Pull; Transformer-Coupled Push-Pull Power Amplifier.*

Push-Pull Frequency Multiplier
(Figure P-40)

In this circuit the input tank is tuned to the frequency of the fundamental signal, and the output tank is tuned to the desired harmonic. This must be an odd harmonic, since even harmonics are canceled out by the push-pull action. The two transistors are connected in a common-emitter configuration, and are biased for Class-C operation during each cycle of the fundamental frequency. Unless the input drive level is high enough, a small forward bias voltage must be provided. See *Frequency Conversion Circuits.*

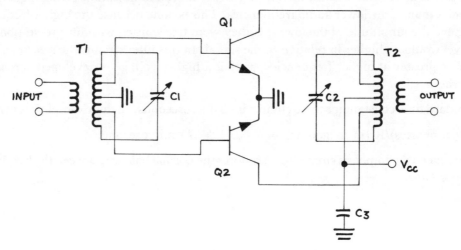

Figure P–40
Push-Pull Frequency Multiplier

Parts List

C1, T1	Input tank, tuned to fundamental frequency
C2, T2	Output tank, tuned to desired harmonic
C3	0.02 microfarad
Q1, Q2	GE11 or equivalent (should be matched pair)

Push-Pull Frequency Multiplier
(Figure P-41)

This circuit closely resembles a push-pull Class-C amplifier, except that neutralization capacitors are not required, since the output frequency is not the same as the input frequency. The input tank is tuned to the fundamental frequency, and the output tank is tuned to the desired harmonic. This must be an odd harmonic, since even harmonics are canceled out by the push-pull action. The highest obtainable harmonic is largely determined by the bias applied to the tubes. They are biased for Class-C operation with the conduction angle of each tube considerably less than 180 degrees. The higher the multiplication factor, the narrower the conduction angle must be for high efficiency. See *Frequency Conversion Circuits.*

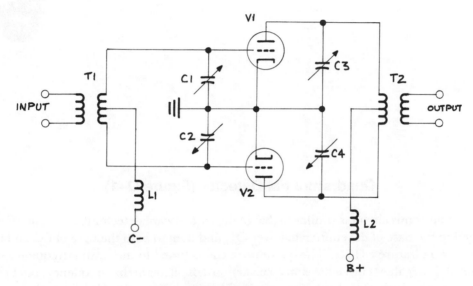

Figure P–41
Push-Pull Frequency Multiplier

Parts List

C1, C2, T1	Input tank, tuned to fundamental frequency
C3, C4, T2	Output tank, tuned to desired harmonic
L1, L2	RF chokes, 2.5 microhenries
V1, V2	12AU7 (twin triode) or equivalent

Quadrature FM Detector (Figure Q–1)

The principle of this circuit is similar to that of the gated-beam detector (q.v.). The IF signal is applied to the base of the emitter-follower Q1, and then to both the base of Q2 and to the tap on the quadrature coil L1. The quadrature coil is tuned to the resting frequency of the IF signal (4.5 megahertz in a television receiver), so it oscillates at that frequency, but because of the method of excitation the voltage across C4 is 90 degrees out of phase with the signal voltage on the base of Q2. The output signal from Q2 is applied between the emitter and base of Q4, so the situation now is that Q4 is receiving an inphase signal of 4.5 megahertz, while Q3 is receiving a 4.5-megahertz signal phase shifted 90 degrees.

Q3 and Q4 are biased so that each is turned on by the positive peaks of the IF signal, but due to the phase shift they are only *both* on at the same time for a fraction of a complete cycle. During this time the current in R5 is greater than what it would be if only one transistor were conducting. This changes the bias on Q4, so that Q4 conducts more heavily.

When the FM signal is modulated, its frequency increases and decreases in accordance with the amplitude of the audio signal with which it is modulated. Since C4-L1 are tuned to the resting frequency, the variations in frequency cause the resonant circuit to fluctuate between

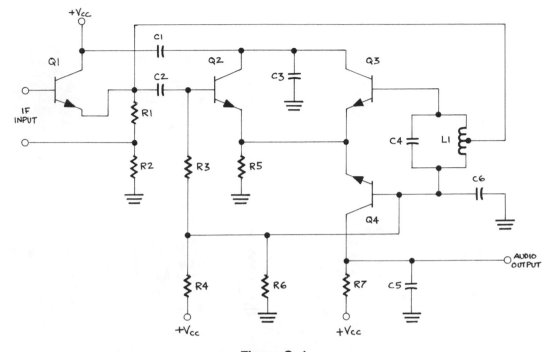

Figure Q–1
Quadrature FM Detector

being partly inductive and partly capacitive, in addition to being purely resistive at resonance. These fluctuations in reactance cause the phase of the signal on Q3's base to shift with respect to that on Q4's base. As a result, the amount of overlap when both transistors are conducting varies, and this in turn causes the duration of increased current through R7 to vary. Consequently, the amplitude of the charge on C5 varies also at the audio rate of the modulating signal.

The circuit has a disadvantage, in that a quadrature coil is required to provide the 90-degree phase shift. This means that the entire circuit cannot be provided on an IC. Newer ICs, such as the MC1358/CA3065, provide everything on a 14-pin dual in-line package (DIP). See *Demodulation Circuits*.

Quadrature Oscillator (Figure Q–2)

This circuit uses two operational amplifiers to generate a sine and cosine output consisting of two sine waves with a frequency of about 1 kilohertz, phase-shifted 90 degrees with respect to each other. See *Oscillator and Signal Generator Circuits*.

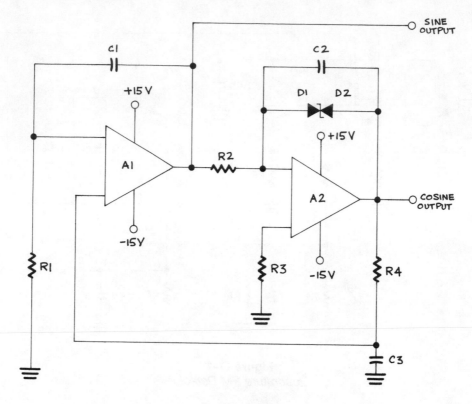

Figure Q–2
Quadrature Oscillator

Parts List

A1, A2	Operational amplifiers, 741C, or equivalent
C1, C2, C3	820 picofarads, 1 percent
D1, D2	1N4733 or equivalent
R1	180 kilohms, 1 percent
R2, R3, R4	190 kilohms, 1 percent

Radiotelegraph Transmitter (Figure R–1)

In this circuit, Q1 is the master oscillator and Q2 is the power amplifier. The meter M1 indicates Q2 collector current, used to obtain optimum tuning of the tank circuit L2-C2 and maximum loading of L3. Both transistors are turned on and off by the telegraph key when transmitting code signals. Maximum power input to the final is 5 watts. See *Oscillator and Signal Generator Circuits*.

Parts List

C1	0.001 microfarad
C2	500 picofarads
C3	1000 picofarads
C4, C5	Variable capacitors for tuning oscillator tank circuit
C6	100 picofarads
C7, C8	0.01 microfarad
C9	0.001 microfarad
C10	Variable capacitor for tuning output tank circuit
C11	0.001 microfarad
L1	2.5 millihenries, RF choke
L2, L3	Oscillator and output tank tuning coils, according to frequency
M1	0-1 mA, DC panel meter
Q1	GE11 or equivalent
Q2	GE277 or equivalent

R1   120 kilohms
R2   100 kilohms
R3, R4   2.2 kilohms
S1   Telegraph key

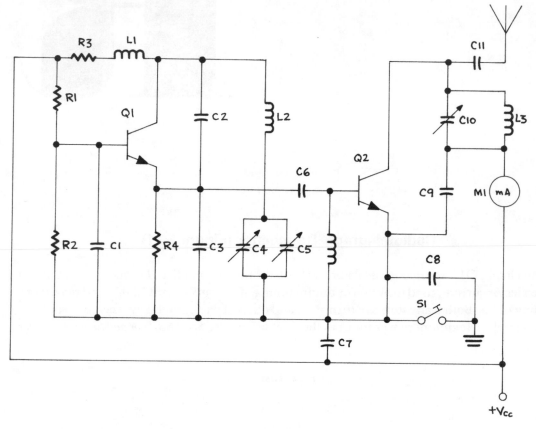

Figure R–1
Radiotelegraph Transmitter

Ratio Detector (Figure R–2)

The ratio detector is widely used in FM receivers to recover the audio from the FM signal. As in all FM detectors, its purpose is to convert the frequency deviations of the FM signal to amplitude variations corresponding to the original modulating signal. The ratio detector also does away with the need for a separate limiting stage.

There are three windings in the transformer T1. The IF signal from the preceding stage is applied to L1, from which it is coupled inductively to L2 and L3. The signal appearing in L3 is in phase with the signal in L1, but that in L2 is 90 degrees different in phase.

When the incoming signal is at the resting (unmodulated) frequency of 10.7 megahertz, to which the windings are tuned, the vector sum of the voltage across the upper half of L2 and the voltage across L3 is equal to the vector sum of the voltage across the lower half of L2 and the voltage across L3. Therefore the currents flowing through each diode are equal, and the capacitors C4 and C5 are charged equally. But when the frequency of the incoming signal deviates as a result of modulation, the vector sums are no longer equal because the signal is above or below the resonant frequency, so that the phase angle alters. The currents through the diodes are now unequal, their inequality being proportional to the frequency deviation. C4 and C5 therefore charge to different voltages.

These voltages vary at the rate of the modulation frequency. In other words, they vary at an audio frequency. At this much lower frequency the inductances of L2 and L3 are negligible, as if they were straight wires, so the voltage appearing at the centertap of L2 will be according to the *ratio* between the voltages on C4 and C5.

The electrolytic capacitor C6, together with R3 and R4, forms a stabilizing circuit across the detector output. The current flow through the circuit will charge C6 to the average level of the incoming signal. Any undesired variations or impulse noise above this average value will have no effect upon the audio output because the voltage on C6 cannot change fast enough to follow the voltage variation. Thus the circuit acts as its own limiter. See *Demodulation Circuits*.

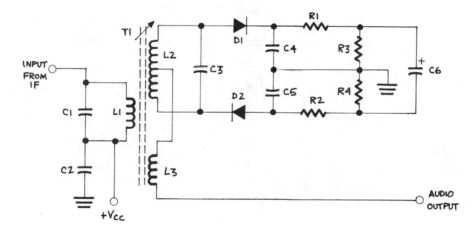

Figure R–2
Ratio Detector

Parts List

C2	0.01 microfarad
C4, C5	270 picofarads
C6	5 microfarads, 25 volts
D1, D2	1N64 or equivalent

R1	1.5 kilohms
R2	1 kilohm
R3, R4	6.8 kilohms
T1	Ratio-detector transformer consisting of tuned circuits L1-C1, L2-C3, and L3 (TRW23148, etc.)

RC-Coupled Amplifiers (Figure R–3)

RC means resistance-capacitance, and this form of interstage coupling is the one most commonly used for vacuum-tube audio amplifiers. The reason for this is that the capacitor, C1 in (A),

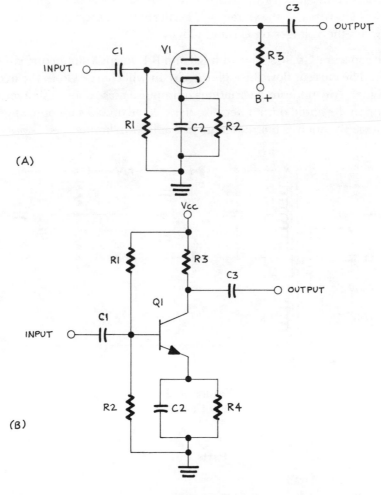

Figure R–3
RC-Coupled Amplifiers

provides isolation from any DC component that might otherwise be coupled into the input stage from the output of the preceding stage. Typical values for these components are 0.05 microfarad and 500 kilohms. These values ensure that little signal voltage is dropped across C1, and practically all of it appears across R1. The resistor is often variable, for use as a level or volume control.

Unlike the vacuum tube, a transistor in a common-emitter circuit has a rather low input impedance, which would shunt a high-value resistor. Also, this resistor is important in establishing the bias voltage on the base of the transistor, so its value is determined from this point of view. The value of C1 must be made considerably higher than would be the case for a vacuum-tube circuit, because the resistance between the base and ground is low, and appreciable reactance in the capacitor would make it too large a component in the voltage divider consisting of C1 and R2 paralleled by Q1 and R4. A typical value for C1 would therefore be 1 microfarad. See *Amplifier Circuits*.

RC Oscillator (Figure R–4)

An RC oscillator is one in which the oscillation frequency is generated by the time constant of capacitors and resistors in the circuit. The time constant of the resistor and capacitor in Figure R-4 is given by:

$$T = RC = (5 \times 10^4 \text{ ohms}) (1 \times 10^{-6} \text{ farad})$$
$$= 0.05 \text{ second}$$

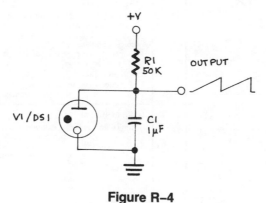

Figure R–4
RC Oscillator

In this circuit DC flows through the resistor to charge the capacitor. The potential across the capacitor is also across the gas regulator tube. When the potential reaches 120 volts the tube fires, discharging the capacitor almost instantaneously until its potential is down to about 80 volts, at which level the tube ceases to conduct. The capacitor now recharges slowly through the resistor until it reaches 120 volts again, when the cycle is repeated.

We want to find the frequency of the sawtooth signal that this circuit generates. This signal consists of two portions, a slow rise and a fast fall. Since the fall time is negligible compared to the rise time, we shall ignore it. The rise time is obtained in the following way.

The maximum capacitor potential is 120 volts and the minimum 80 volts. Therefore we want to find the interval needed for the capacitor potential to rise from 80 volts to 120 volts. The lower voltage is 80/120 = 0.67 times the upper. Referring to Figure R-5, we locate the point on the vertical scale corresponding to 0.67 times the maximum voltage. Then we move horizontally from this point until we encounter curve 1. The corresponding value on the horizontal scale is 1.15. Thus the capacitor potential rises to 120 volts after 1.15 time-constant units. The time constant we have already determined to be 0.05 second, so the interval needed for the capacitor potential to rise from 80 to 120 volts is:

$$t = 1.15 \times 0.05 = 0.0575 \text{ second}$$

Since we are neglecting the fall time, this is the duration of one cycle of the wave; therefore the frequency is given by its reciprocal:

$$f = 1/t = 1/0.0575 = 17.4 \text{ hertz}$$

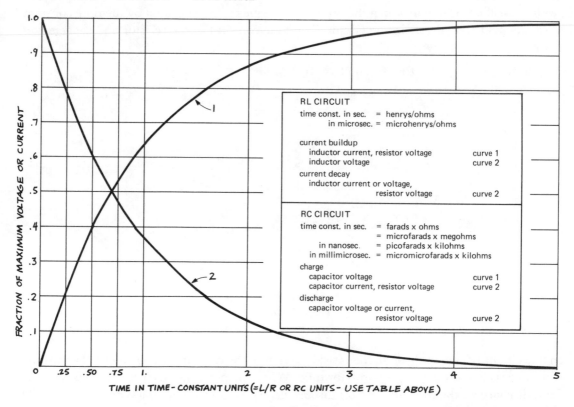

Figure R–5
Time Constant Curves

Most RC oscillators are relaxation oscillators, like the one just described, and generate pulse, sawtooth, square-wave, and other nonsine-wave signals. However, there are also sine-wave RC oscillators, such as the Wien-bridge oscillator (q.v.), which operate at audio frequencies. See *Oscillator and Signal Generator Circuits.*

RCTL NOR Gate (Figure R–6)

RCTL is the abbreviation for "resistor-capacitor-transistor logic." This gate was designed to try and improve the switching speed of the basic NOR gate (q.v.), but has been superseded by TTL and ECL gates because the latter do not require the added complication and expense of providing capacitors in an integrated circuit. See *Logic Circuits.*

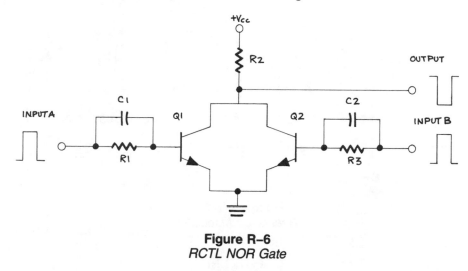

Figure R–6
RCTL NOR Gate

Parts List

C1, C2	100 picofarads
Q1, Q2	GE20 or equivalent
R1, R3	470 ohms
R2	640 ohms

Reactance Modulator (Figure R–7)

The simplest and most satisfactory device for amateur FM is the reactance modulator. This is a vacuum tube connected to the RF tank circuit of an oscillator in such a way as to act as a variable inductance or capacitance. In the diagram, the control grid of V2 is connected across the small capacitance of C8, which represents the input capacitance of the tube, in series with R3 across the oscillator tank circuit. Any type of oscillator circuit may be used. The resistance of R3 is made large compared to the reactance of C6, so the RF current through R3-C6 will be practically in phase with the RF voltage appearing at the terminals of the tank circuit.

However, the voltage across C6 will lag the current by 90 degrees. The RF current in the plate circuit of V2 will be in phase with the grid voltage, and consequently is 90 degrees behind the current through C6, or 90 degrees behind the RF tank voltage. This lagging current is drawn through the oscillator tank, giving the same effect as though an inductance were connected across the tank. The frequency increases according to the amplitude of the lagging plate current of V2. The value of this is determined by the voltage on the control grid of V2. Hence the oscillator frequency will vary when an audio signal is applied to the control grid. See *Modulation Circuits.*

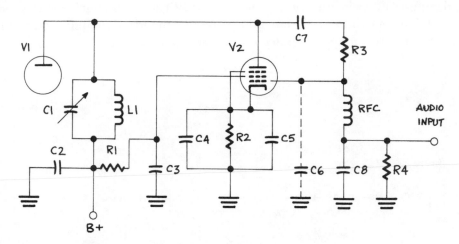

Figure R–7
Reactance Modulator

Parts List

C1	RF tank capacitance, according to frequency
C2, C7	0.001 microfarad
C3, C5, C8	0.0047 microfarad
C4	10 microfarads, electrolytic
C6	Tube input capacitance (if necessary, a 3-30 picofarad trimmer capacitor may be connected here to control modulator sensitivity)
L1	RF tank inductance
R1	1.2 megohms
R2	1.5 kilohms
R3, R4	470 kilohms
RFC	RF choke, 2.5 millihenries
V1	Tube of previous oscillator stage
V2	6BJ6 or equivalent

Reflex Amplifier (Figure R–8)

This circuit uses one transistor for both an IF stage and an audio amplifier. The IF signal is coupled from the previous stage by T1 to the base of Q1, and after amplification is coupled out from the collector via T2 to the diode detector D1. R5 is both the diode load and the volume control.

The audio signal is fed back via C1 to the point denoted AUDIO IN, from which it is connected to the base of Q1 via the secondary of T1 (which has negligible reactance at audio frequencies). After amplification by Q1 it appears across R4 (the primary of T2 also has negligible reactance to an audio signal), and is coupled to the audio output amplifier via C5. See *Amplifier Circuits*.

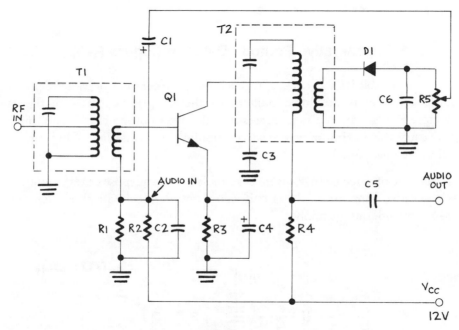

Figure R–8
Reflex Amplifier

Parts List

C1, C5	10 microfarads
C2, C3	0.02 microfarad
C4	100 microfarads
C6	0.001 microfarad
D1	1N34A or equivalent
Q1	GE20 or equivalent

R1	2.2 kilohms
R2	15 kilohms
R3	2.5 kilohms
R4	2.2 kilohms
R5	2.5 kilohms, variable
T1	IF interstage transformer
T2	IF output transformer

Regenerative JFET Detector

See *FET Regenerative Detector*.

Regenerative Pentode Detector (Figure R–9)

By providing controllable RF feedback, or regeneration, in a pentode (or triode) detector circuit, the incoming signal can be amplified many times, thereby greatly increasing the sensitivity of the detector. Regeneration also increases the effective Q of the circuit and improves the selectivity because the maximum regenerative amplification takes place only at the frequency to which the circuit is tuned.

The grid-leak type of detector (q.v.) is most suitable for the purpose of regenerating. Except for the regenerative connection, the circuit values are identical with those described for it, and the same considerations apply.

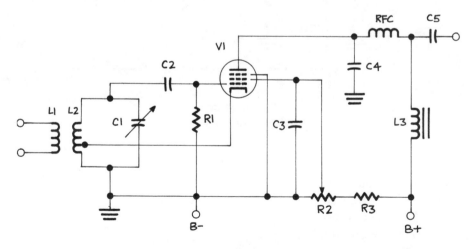

Figure R–9
Regenerative Pentode Detector

The amount of regeneration must be controllable, because maximum regenerative amplification is secured at the critical point where the circuit is just about to oscillate, and the critical point in turn depends upon circuit conditions, which may vary with the frequency to which the detector is tuned. At this critical point the circuit can be made to oscillate by body capacitance, a condition that was taken advantage of in early British aircraft receivers, where a connection to the grid circuit was brought out to a small metal stud on the front panel, so that the operator could, by touching it, determine if the receiver was close to the oscillating point. Radio operators were consequently known as "lickers and tappers," since they were constantly licking a finger and tapping this stud while adjusting the regeneration control.

In this circuit, regeneration is controlled by adjusting the screen voltage. The oscillating circuit is of the Hartley type, and since the screen and plate are in parallel for RF only a small amount of "tickler"—that is, relatively few turns between the cathode tap and ground—is required for oscillation. The number of turns should be varied until V1 just goes into oscillation when the screen grid is at about 30 volts. The portion of R2 between the sliding contact and ground is bypassed by C3 to filter out the scratching noise when the arm is rotated. See *Demodulation Circuits*.

Parts List

C1-L2	Input circuit, tuned to the signal frequency
C2, C4	100 picofarads
C3	1 microfarad
C5	0.1 microfarad
L3	500 henries
R1	1 to 5 megohms
R2	50 kilohms, variable
R3	50 to 100 kilohms
V1	6BA6 or equivalent

Relaxation Oscillator

See *RC Oscillator*.

Remote-Control Transmitter

See *Tunnel-Diode AM Transmitter*.

Repeater Coil (Figure R–10)

This is the name given by the telephone company to a special audio-frequency transformer that has four windings which can be connected in various ways. In (A), both the primary and secondary windings are connected in series, which gives a 600-ohm impedance in both directions. In (B), the secondary windings are connected in parallel, which gives an impedance of 150 ohms, allowing a 150-ohm source to be matched to a 600-ohm load. Thus, this transformer can be used to match the various impedances the telephone company is interested in. It also can be used to block DC while transmitting AC. See *Passive Filters.*

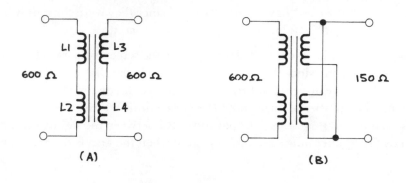

Figure R–10
Repeater Coil

Resistance-Capacitor-Coupled Transistor Amplifier

See *RC-Coupled Amplifiers.*

Resistance-Capacitor-Coupled Triode Amplifier

See *RC-Coupled Amplifiers.*

Resistor-Input Filter

See *Passive Filter Circuits; Power Supply Circuits.*

Resonance Bridge (Figure R–11)

When this bridge is balanced the bridge arm consisting of R_x, C_1, and L_x is resonant at the frequency of the applied signal. At resonance the reactances of C_1 and L_x are equal and opposite, and cancel each other, so that the circuit is purely resistive. This simplifies measurement of inductances and impedances. An inductive value is given by:

$$L_x = 1/6.28fC_1$$

The resistive component is given by:

$$R_x = R_sR_1/R_2$$

See *Test and Indicating Circuits*.

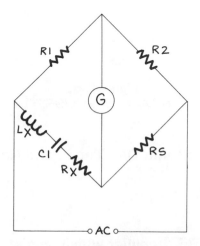

Figure R–11
Resonance Bridge

RF Amplifier with IGFET (Figure R–12)

Since an insulated-gate field-effect transistor (IGFET) is very similar to a triode tube in its behavior, neutralization is required. This is obtained by C3, which provides sufficient external feedback to neutralize the feedback which occurs internally in the transistor. See *Amplifier Circuits*.

Parts List

C1-T1	Input circuit tuned to RF signal
C2, C4	0.05 microfarad
C3	1-10 picofarads neutralizing capacitor
C5-T2	Output circuit tuned to RF signal
Q1	N-channel IGFET (depletion type)
R1	330 ohms
R2	560 ohms

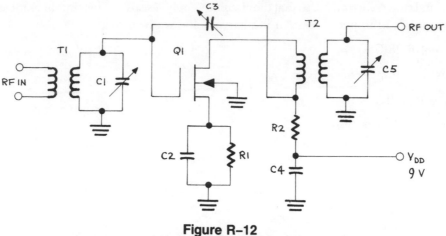

Figure R–12
RF Amplifier with IGFET

RF Amplifier with One Triode and Neutralization

See *Neutralized Triode RF Amplifier.*

RF Amplifier with Transistor

See *Neutralized Transistor Amplifier.*

RF Amplifier with Two Triodes in Push-Pull

See *Neutralized Push-Pull Triode RF Amplifier.*

RF Output Meter (Figure R–13)

In low-power radio transmitters it is possible to measure the relative power output with a milliammeter. In Diagram R-13, a sample of the signal is fed to the meter circuit via C3. It is then rectified by D1 and applied to R1 and M1. C2 bypasses RF and stabilizes the meter. This circuit can be calibrated by measuring actual RF power with an RF wattmeter connected to the antenna jack, and adjusting C3 to obtain equivalent readings on M1. See *Test and Indicating Circuits.*

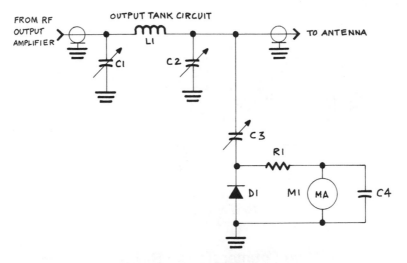

Figure R–13
RF Output Meter

Parts List

C1, C2, L1	Output tank circuit of transmitter
C3	35 picofarads, variable
C4	0.01 microfarad
D1	1N34A or equivalent
M1	0-1 milliamperes, DC, or as available
R1	Multiplier resistor as required for meter

RF Power Amplifier with Two Triodes in Push-Pull

See *Neutralized Push-Pull Triode RF Amplifier.*

Rice Neutralization (Figure R–14)

See *Neutralized Triode RF Amplifier.* The only difference between the two circuits is the manner of connecting the feedback. See *Amplifier Circuits.*

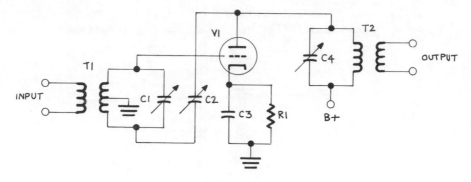

Figure R–14
Rice Neutralization

Ring Counter (Figure R–15)

A ring counter is a serial shift register in which the output of the last flip-flop is fed back to the input of the first, so that data are not only shifted but recirculated as well. Thus, with four stages a given pattern is repeated in response to every fourth clock pulse. See *Logic Circuits*.

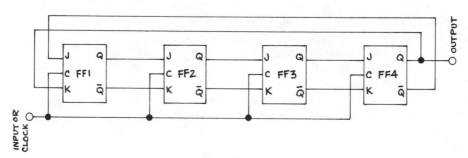

Figure R–15
Ring Counter

Ring Modulator (Figure R–16)

In this circuit four diodes function as if they were a double-pole, double-throw switch, operating at the carrier frequency. The carrier is applied to the centertaps of T1 and T2, so there is no carrier output in the absence of modulation. When a modulating signal is applied upper and lower sidebands are generated. This circuit is therefore used to generate a double-sideband, suppressed-carrier signal. See *Modulation Circuits*.

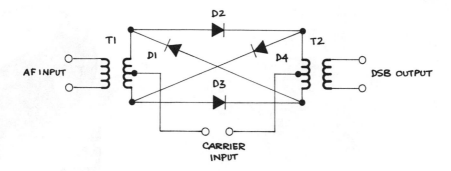

Figure R–16
Ring Modulator

RTL NAND Gage

See *NAND Gate*.

RTL NOR Gate

See *NOR Gate*.

Saturated Flip-Flop

See *Bistable Multivibrator*.

Sawtooth Generator (Figure S-1)

A sawtooth signal has a waveshape consisting of a linearly rising voltage, followed by a linearly falling voltage, or vice versa. If their slopes were equal they would form a triangle wave, but in a sawtooth one slope is relatively gradual compared to the other, which is very steep, almost vertical.

This waveshape is achieved by charging a capacitor through a resistor until the voltage across the capacitor reaches a value that triggers some device into discharging it practically instantaneously, as was described in some detail under *RC Oscillator* (q.v.). This circuit is very similar, with a unijunction transistor (UJT) replacing the gas regulator tube. The UJT turns on when its emitter reaches 11 volts. At this point the charge on the capacitor is shorted to ground by the very low resistance between the emitter and base 2.

Such a sawtooth is not very linear, because the capacitor does not charge linearly. As the voltage on the capacitor builds up it opposes the charging voltage more and more, so that the voltage across the resistor diminishes. This leads to a gradual reduction in the current flowing into the capacitor. If a linear sawtooth is required, as in an oscilloscope sweep, a more complex circuit, such as a *phantastron* (q.v.), has to be used. See *Oscillator and Signal Generator Circuits*.

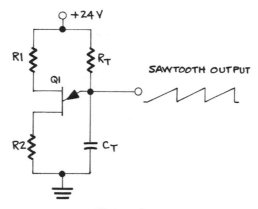

Figure S–1
Sawtooth Generator

Parts List

C_T, R_T Values to be selected for the desired frequency (see under *RC Oscillator*)
Q1   HEP310 or equivalent
R1   1 kilohm
R2   200 ohms

Scaling Inverter with Operational Amplifier (Figure S-2)

By connecting an op-amp in this way a precise ratio between the output voltage and the input voltage is given by:

$$V_{out} = V_{in}(R_f/R_i)$$

The output is also inverted with respect to the input. The op-amp can be the popular 741 type, and the resistor values will depend upon the desired scale factor. See *Amplifier Circuits; Operational Amplifier Circuits; Logic Circuits.*

Schering Bridge (Figure S-3)

The Schering bridge is useful in higher-voltage applications. A constant voltage is assured across the capacitor being measured (C_X). When the bridge has been balanced this value is given by:

$$C_X = C_S(R_2/R_1)$$

See *Test and Indicating Circuits.*

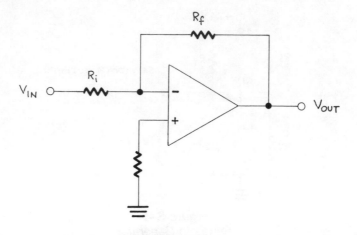

Figure S–2
Scaling Inverter with Operational Amplifier

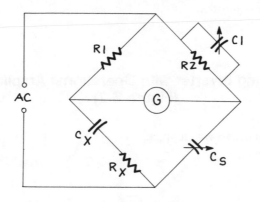

Figure S–3
Schering Bridge

Schmitt Trigger

See *Monostable Multivibrator.*

Selectivity Filter

See *IF Amplifier with Ceramic Filter.*

Series and Parallel Resonant Wavetraps (Figure S-4)

A series resonant wavetrap is shown at (A). The inductor or the capacitor may be variable, with values that allow the circuit to be resonant at the frequency of an interfering signal. It will then have a very low impedance, grounding the signal. At all other frequencies it will have a high impedance, with little or no effect upon desired frequencies.

A parallel resonant wavetrap is shown at (B). Its operation is opposite to that of the series trap. When tuned to the frequency of the interfering signal it offers a very high impedance to it, thereby blocking it, but it offers a low impedance to other frequencies. See *Antenna Circuits*.

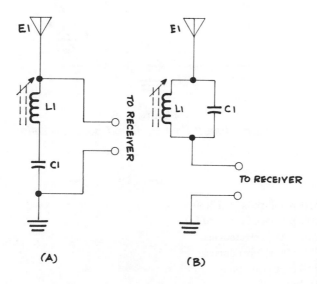

Figure S–4
Series and Parallel Resonant Wavetraps

Series-Tuned Emitter, Tuned-Base Oscillator (Figure S-5)

This oscillator depends upon feedback per the emitter-base junction capacitance to sustain oscillation. Both tuned circuits are resonant at the same frequency, but it is necessary for the reactance of C2 to be slightly larger than the reactance of L2 to provide a leading reactance. This gives stability without the use of elaborate nonlinear components. See *Oscillator and Signal Generator Circuits*.

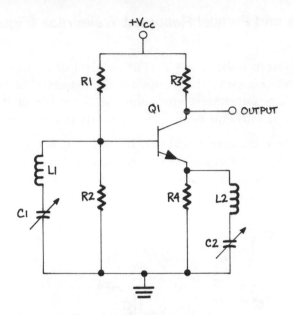

Figure S–5
Series Tuned Emitter, Tuned Base Oscillator

Parts List

C1	510 picofarads, variable
C2	100 picofarads, variable
L1	200-500 microhenries
L2	300-500 microhenries
Q1	2N78 or equivalent
R1, R2,	10 kilohms
R4	
R3	5 kilohms

Series Voltage Regulator with Comparator
(Figure S-6)

In this voltage regulator circuit, Q3, Q5, and Q7 are connected in parallel, and act as a variable resistance between the unregulated DC input and the regulated DC output. The output voltage divider R8-R9-R10 enables a sample of the output voltage to be applied to the base of Q6. With Q4, this transistor forms a comparator that senses any change in the output voltage with reference to the zener diode D1, and applies the resultant change in Q4's collector voltage to the base of Q1, the first transistor in the pair consisting of Q1 and Q2.

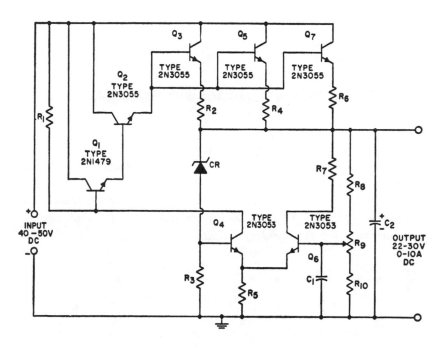

Figure S–6
Series Voltage Regulator with Comparator

The output voltage from this pair is applied to the bases of all three parallel transistors Q3, Q5, and Q7. Their conduction varies in accordance with this bias, increasing when the output voltage has a tendency to fall, and decreasing when it is inclined to rise. This counteracts variations in both load current and line potential. See *Power Supply Circuits.*

Parts List

C1	0.01 microfarad
C2	20 microfarads
D1	1N4744 (15 V) or equivalent
Q1	2N1479 or equivalent
Q2, Q3, Q5, Q7	2N3055 or equivalent
Q4, Q6	2N3053 or equivalent
R1	4.7 kilohms
R2, R4, R6	0.1 ohm
R3	390 ohms

R5	560 ohms
R7	3.9 kilohms
R8	3.3 kilohms
R9	500 ohms, variable
R10	2.2 kilohms

Series Voltage Regulator with Preregulator
(Figure S-7)

The series voltage regulator can be refined by adding a preregulator. This consists of the components inside the dashed line. D1 maintains a constant voltage across R2, and therefore a constant emitter current in Q1. Through the comparison circuit, Q3 takes a lesser or larger percentage of the current from the constant-current source, thus controlling the base current of Q2 more effectively. Q2 and Q4 form a Darlington pair, which provides greater gain than a single series transistor. See *Power Supply Circuits*.

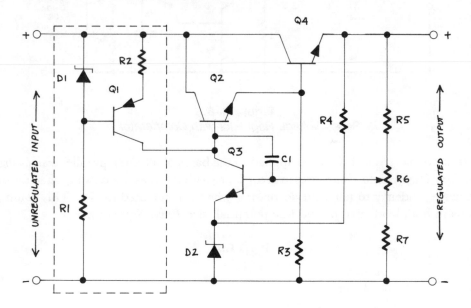

Figure S–7
Series Voltage Regulator with Preregulator

Parts List

C1	0.01 microfarad
D1	1N746 or equivalent
D2	1N753 or equivalent
Q1	2N1131 or equivalent

Q2	2N656 or equivalent
Q3	2N338 or equivalent
Q4	2N1049 or equivalent
R1	5.1 kilohms
R2, R7	1 kilohm
R3	30 kilohms
R4	6.2 kilohms
R5	4.7 kilohms
R6	500 ohms, variable
R7	1 kilohm

Shunt Degeneration (Figure S-8)

Shunt degeneration or negative feedback is used to stabilize a transistor circuit for slow or DC variations. In this configuration any increase in collector current causes the collector voltage to decrease, so that the voltage output of the bias voltage divider also decreases. A reduced base bias voltage lowers the base current and tends to oppose the original rise in collector current. A similar reaction in the opposite direction takes place when the collector current decreases.

This method is seldom sufficient by itself, however. An emitter resistor is generally used in addition to provide negative feedback for slow variations, as caused by heating. A slow change in emitter current is counteracted by the resultant emitter-base voltage change across the emitter resistor. A capacitor is connected in parallel with this resistor to reduce this negative feedback for AC voltages. The complete circuit is analyzed for design purposes under *Amplifier Circuits*.

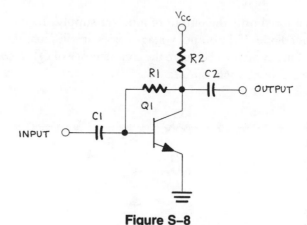

Figure S–8
Shunt Degeneration

Shunt Diode Clipper

See *Clipper*.

Shunt-Fed Triode Power Amplifier (Figure S-9)

When it is necessary to keep the plate voltage out of the load circuit it may be supplied to the plate via an RF choke. The inductance of this choke must be high compared to the load impedance, so that negligible RF is grounded through the power supply. A commonly used value for VHF transmitters is 2.5 millihenries. This choke must also be able to handle the DC plate current. The resistor R1 merely symbolizes the load, which may be either a parallel-resonant transformer or a pi-network. See *Amplifier Circuits*.

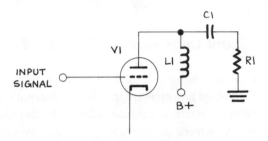

Figure S–9
Shunt-Fed Triode Power Amplifier

Shunt Voltage Regulator (Figure S-10)

In this circuit a transistor shunts the output of a power supply. Its base voltage is held to a fixed value by the zener diode. If the output voltage varies, it will cause the emitter-base voltage to vary accordingly. This in turn will change the conductance of Q1 to counteract the change in output voltage. See *Power Supply Circuits*.

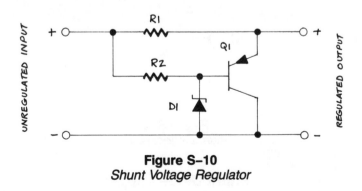

Figure S–10
Shunt Voltage Regulator

Parts List

D1   Zener diode with zener voltage according to desired output voltage. (For instance, if zener voltage is 4.3, output voltage will be 5.0 volts.)

Q1   GE84 or equivalent

R1   180 ohms, 2 watts

R2   1 kilohm

Simplexed-to-Ground Control (Figure S-11)

This is another example of the use of telephone repeater coils (q.v.), one in which a DC control signal can be sent over the same pair of wires that are used for speech transmission. When S1 is closed, DC flows through both wires in the same direction to actuate the relay K1 (the DC return path is through the ground). The DC has no effect on speech transmission sharing the same wires.

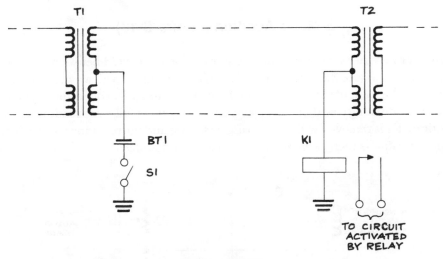

Figure S–11
Simplexed-to-Ground Control

Single-Circuit Receiver

See *Regenerative Detector*.

Single Series-Resonant Wavetrap (Figure S-12)

Since a series-resonant circuit such as L1 and C1 in the diagram is effectively a short circuit to a signal at the frequency to which it is tuned, it can be used to eliminate an interfering

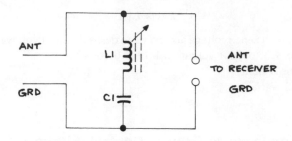

Figure S–12
Single Series-Resonant Wavetrap

signal by connecting it between the antenna and ground terminals of a receiver with an unbalanced input. The values of L1 and C1 will depend upon the frequency of the interfering signal. See *Antenna Circuits.*

S-Meter Addition (Figure S-13)

A milliammeter in series with a suitable resistor can be connected between any convenient point on the AVC bus (see *Automatic Volume Control*) and ground to give a meter that indicates signal strength. The value of the resistor is best found by temporarily connecting a potentiometer in its place, and adjusting it so that a strong local station gives a full-scale reading on the meter. The potentiometer is removed without altering its setting and the resistance measured. A fixed resistor of this value is then permanently connected in its place.

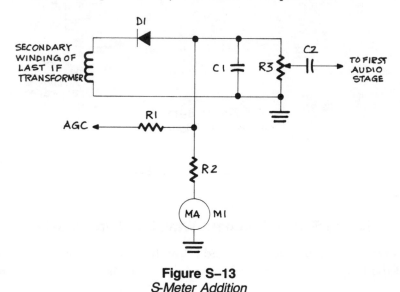

Figure S–13
S-Meter Addition

The milliammeter may also be rescaled in S units. The S-unit scale has 6-decibel steps from 0 to 9, with S9 the highest. (S9 is the value corresponding to a signal strength of 100 microvolts.) If an S meter is available instead of a milliammeter, it may be calibrated in the same way. See *Test and Indicating Circuits*.

S-Meter Bridge

See *Bridged S-Meter*.

Solid-State Detector and AVC (Figure S-14)

T1 is the output IF transformer, with its primary winding tuned by C1. The amplified IF signal is applied to D1, the detected signal appearing across the diode load resistor and volume control R2. C2, R1, and C3 filter out the IF components of the detected signal and provide AVC voltage. In this circuit it is negative, but a positive voltage may be required for control of stages with PNP transistors. See *Demodulation Circuits*.

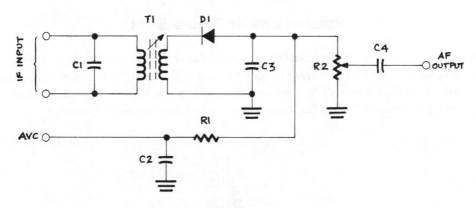

Figure S–14
Solid-State Detector and AVC

Parts List

C1-T1	IF transformer with primary tuned by C1
C2	0.01 microfarad
C3, C4	10 microfarads
D1	1N34 or equivalent
R1	1.8 kilohms
R2	2.5 kilohms, variable

Speaker Combiner (Figure S-15)

When you want to add an extension speaker to your stereo, you need this circuit for the sound of both channels. The circuit uses a three-winding transformer, such as the Mix-n-Match (Alco Electronic Products Co.), which has two primary windings and one secondary winding. The transformer presents an 8-ohm impedance both ways. See *Amplifier Circuits.*

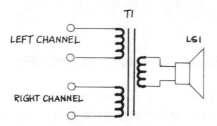

Figure S–15
Speaker Combiner

Speaker Splitter (Figure S-16)

The three-winding transformer used as a speaker combiner (q.v.) can also connect two speakers to the same source. Although a transformer such as this is seldom necessary (the speakers are usually connected directly), it may be the answer to an impedance problem, since the transformer presents an 8-ohm impedance both ways. See *Amplifier Circuits.*

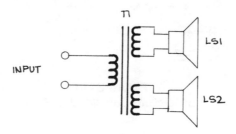

Figure S–16
Speaker Splitter

Squelch (Figure S-17)

Before this circuit was invented the operator of a communications radio station had to put up with a continuous roar from his receiver when no signal was being received. In the absence of a signal, there was no AVC voltage (q.v.) to reduce the sensitivity of the receiver, so it would run flat out, amplifying noise and other interference to the maximum degree.

Squelch is a means of killing the audio amplifier between signals, but automatically restoring it to operation when a signal is received. A squelch control enables the operator to set the level of signal that will do this. As shown in the circuit, a squelch tube V1 is conducting heavily in the absence of AVC voltage (which is generated only by an incoming signal). The plate current causes a considerable voltage drop across R1, so that the plate voltage is low. This low voltage is applied to the grid of the audio amplifier V2, keeping it cut off.

When a signal is received, the negative AVC voltage it generates is coupled to the grid of V1. If it is negative enough to turn the tube off, V1's plate voltage rises to a level that allows V2 to conduct, and it resumes amplifying the audio signal applied to its grid. The level of AVC voltage required to turn V1 off depends upon the setting of the squelch control. See *Demodulation Circuits.*

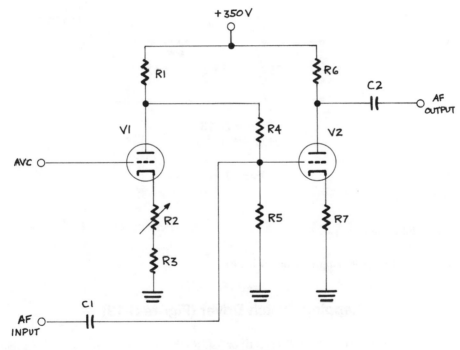

Figure S–17
Squelch

Parts List

C1, C2	0.05 microfarad
R1	25 kilohms
R2	1 kilohm, variable ("squelch control")
R3	100 ohms
R4, R5	0.5 megohm

R6	5 kilohms
R7	15 kilohms
V1, V2	12AT7

Step Attenuator (Figure S-18)

This is a simple two-section step attenuator. The left-hand section gives 6 decibels of attenuation, the right 12 decibels. See *Attenuator and Pad Circuits*.

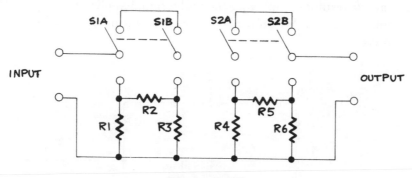

Figure S–18
Step Attenuator

Parts List

R1, R3	150 ohms
R2	33 ohms
R4, R6	82 ohms
R5	91 ohms
S1, S2	Double-pole, single-throw switches

Stepping Switch Driver (Figure S-19)

This circuit consists of a silicon-controlled rectifier (SCR) that is turned on and off by a relaxation oscillator. The AC line input is rectified by D1 and smoothed by C1. The resultant DC charges C2 through R2 and R3. The time taken to charge is adjusted by R2. When the potential on the capacitor reaches about 65 volts the neon lamp DS1 fires, discharging C2; and the momentary current flowing through R5 places a positive voltage on the gate of D2, so that it turns on. Current now flows through the load, which in this case is a stepping switch. However, it flows only for a very short time—just enough to actuate the switch—before the potential on C2, which is also across D2, falls so low that both D2 and DS1 cease to conduct. C2 then starts to charge again, and the cycle is repeated continuously as long as power is applied to the circuit.

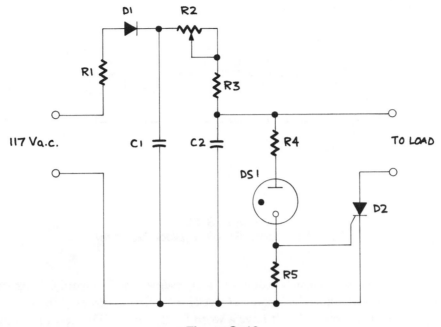

Figure S–19
Stepping Switch Driver

Parts List

C1, C2	25 microfarads
D1	1N4000 series rectifier diode
D2	SCR, 6 amperes, 200 volts, or as required
DS1	NE-2H or equivalent
R1	10 ohms
R2	10 megohms, variable
R3	100 kilohms
R4	10 kilohms
R5	1 kilohm

Superheterodyne Receiver (Figure S-20)

The conventional radio receiver today consists of the sections shown in the block diagram, each of which is described in detail elsewhere in this encyclopedia. The signals picked up by the antenna in most cases are fed directly into the converter circuit (sometimes a separate oscillator and mixer). In television sets, car radios, and some other sets, an RF amplifier precedes the converter stage. In most AM radios the antenna consists of a ferrite-core coil and shunting capacitors, but auto, FM, and TV receivers generally require some type of external antenna for good results.

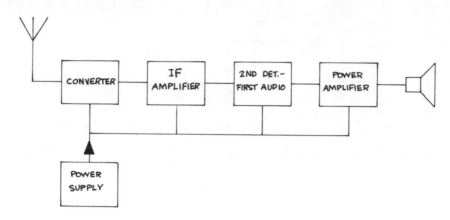

Figure S–20
Superheterodyne Receiver (Block Diagram)

The term *superheterodyne* is a contraction of *supersonic heterodyne*. The first word is Latin for "above sound," or higher than audio. The second is Greek for "different powers," meaning different signal frequencies (the Greeks didn't have a word for that one). This term is an attempt to describe the system of operation in one word, but as that is impossible we'll have to explain it with a few English ones.

The tuning knob of the receiver turns the movable plates of a variable capacitor that has two sections. One section tunes the antenna circuit that selects the signal to be received. The other tunes the local oscillator. The two circuits are designed so that they track. The local oscillator is always tuned to generate a signal an exact number of kilohertz higher than the incoming signal frequency. In AM broadcast reception, this difference is 455 kilohertz as a rule, although other frequencies are possible and have been used. But it is always the same, regardless of the station frequency. For instance, if the AM station has a frequency of 1000 kilohertz, the oscillator generates a signal of 1455 kilohertz. If the AM station frequency is 600 kilohertz, the oscillator frequency is 1055 kilohertz. In every case the difference is 455 kilohertz.

This frequency is called the intermediate frequency (IF) since it is intermediate between RF and AF. It results from mixing the RF and local oscillator signals together. When this is done, two "beat frequencies" are produced, one equal to the sum of the two original frequencies, the other equal to their difference. They all have the same modulation characteristics as the original broadcast signal.

As we want only one of them we couple the output from the converter to the next stage via a tuned transformer that is tuned to the 455-kilohertz IF, which it passes. The other frequencies are blocked. This IF signal is then amplified by the IF amplifier. Although 455 kilohertz is the usual IF for AM broadcasting reception, the IF for FM is 10.7 megahertz, and for television the sound IF is 4.5 megahertz.

Since the IF amplifier has only one frequency to amplify it can be made highly selective, with good gain. It consists of two or more stages, usually coupled by IF transformers, and after the final stage the signal is demodulated in the detector stage. The detector stage is sometimes referred to as the "second detector," to distinguish it from the converter or mixer stage, which used to be known as the "first detector." In the detector the IF carrier frequency is filtered out, leaving only the audio modulation, which is then applied to the audio amplifier. The standard audio amplifier has two stages, a voltage amplifier and a power output amplifier. The latter is a current amplifier. It increases the power of the signal instead of its voltage, so that it can drive a speaker. See *Amplifier Circuits; Antenna Circuits; Demodulation Circuits.*

Suppressed-Carrier (Double-Sideband) Modulator (Figure S-21)

This circuit may be used in a transmitter for double or single sideband radio signals. As shown, it produces both sidebands, but one may be suppressed by the addition of a suitable filter. The audio signal from a microphone or other source is coupled via T1 to the bases of Q1 and Q2, which are connected in push-pull. The amplified audio output from these transistors is coupled by C5 and C7 to the tank circuit L1-C8, but as the inductance of L1 is negligible at audio frequencies no appreciable audio signal voltage appears across it.

The RF carrier is fed to the emitters of Q1 and Q2, so it also appears in the output tank circuit. But since it was not phase split the two signals cancel each other out. In this way the carrier is suppressed.

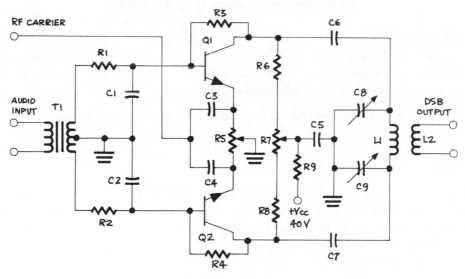

Figure S–21
Suppressed Carrier (Double Sideband) Modulator

At first sight it looks as though we have nothing to radiate, since we have suppressed both the carrier and the audio. However, the mixing of the two frequencies in Q1 and Q2 results in the generation of sum and difference beat frequencies, or, in other words, upper and lower sidebands. For instance, if the carrier frequency is 9 megahertz, and a modulating audio signal of 1 kilohertz is applied, an upper sideband of 9.001 megahertz and a lower sideband of 8.999 megahertz are generated. The tank circuit is resonant to these frequencies, and as they are in push-pull, they appear across L1. See *Modulation Circuits.*

Parts List

C1, C2	0.01 microfarad
C3, C4,	0.001 microfarad
C5, C6,	
C7	
C8-L1	Tank circuit tuned to RF signal frequency
Q1, Q2	GE277 or equivalent
R1, R2	15 kilohms
R3, R4	27 kilohms
R5	30 ohms, variable
R6, R8	5 ohms
R7	50 ohms, variable
R9	25 ohms
T1	Microphone coupling transformer (Thordarson TR-5 or equivalent)

Switching Regulator (Figure S-22)

The switching regulator achieves greater efficiency than other voltage regulators, but at the expense of increased complexity. This disadvantage is overcome to some extent by the use of IC's in the design. However, the one shown in Figure S-22 illustrates discrete components so that its manner of operation can be explained.

Q1 is a switch in this circuit. The current flow from the load via D1 and back to the secondary of T1 is interrupted by this switch, which is turned on and off by a train of square current pulses generated by the voltage comparison circuit Q3 and Q4. The *rate* of these pulses is determined by the values of L1 and C5.

When Q1 is on, current flows from the load through L1, R3, Q1 and D1 to the secondary of T1. In the process it charges C5 to the output voltage level. When Q1 turns off, the diode D3, which previously was reverse biased, starts to conduct. C5 now begins to discharge via L1, R3, and D3, until the decrease in voltage on the base of Q4 drops to below the threshold level established by the zener diode D5.

When this happens, the Schmitt trigger (q.v.) consisting of Q3 and Q4 switches to its other state, terminating the current pulse that was keeping Q1 cut off. The exact moment when

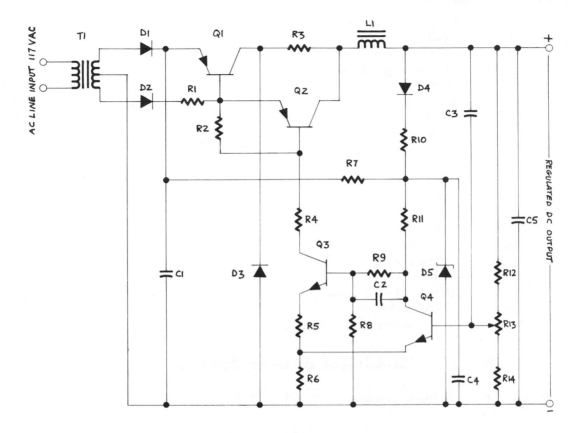

Figure S–22
Switching Regulator

this occurs depends upon how long it takes the voltage to fall below the threshold. If the voltage was high, Q1 will be off longer, and vice versa, so that this tends to keep the voltage constant. See *Power Supply Circuits*.

Parts List

C1	2500 microfarads, 80 volts
C2	470 picofarads
C3	1 microfarad
C4	25 microfarads, 50 volts
C5	5 microfarads, 50 volts
D1, D2	MR1121 or equivalent
D3	1N3880 or equivalent
D4	1N4001 or equivalent

D5	1N4732 or equivalent
L1	3 millihenries (Triad C-49U or equivalent)
Q1	2N3792 or equivalent
Q2	2N4920 or equivalent
Q3	2N4410 or equivalent
Q4	MPS6575 or equivalent
R1	100 ohms
R2	220 ohms
R3	0.2 ohm, 10 watts
R4	1.6 kilohms
R5	24 ohms
R6	1 kilohm
R7	12 kilohms
R8	9.1 kilohms
R9, R12	3.3 kilohms
R10	200 ohms, 2 watts
R11	51 ohms
R13	250 ohms, 1 watt, variable
R14	2.4 kilohms
T1	Power transformer (Triad F-67U or equivalent)

Switch-Type Balanced Modulator

See *Cowan Bridge Balanced Modulator*.

TEST AND INDICATING CIRCUITS

Various test or indicating circuits are located throughout the encyclopedia and are listed at the end of this chapter. The following pages deal with the important subject of meters and how they are used. This may enable you to adapt a standard DC meter for use in a test or indicating circuit.

A neon tester or a pilot light tells us if a circuit is "live," which is a good thing to know before starting work on it. To obtain numerical data (voltage, current, etc.) you need meters. The trouble with meters is that they must use a portion of the energy in the circuit. Where the energy is small, only a small sample can be taken without disturbing the operation of the circuit, which would give a false reading. This is called *loading* the circuit. Figure T-1 shows what happens when a voltmeter with a DC resistance of 10 kilohms is used to measure the voltage across a 10-kilohm resistor. Before connecting the meter the current in the circuit is given by 100 volts/25 kilohms = 4 milliamperes. After the meter has been connected the current in the circuit increases, since the meter and R2 in parallel are only 5 kilohms. The current now is 100 volts/20 kilohms = 5 milliamperes. The voltage across R2 is therefore 5 kilohms × 5 milliamperes = 25 volts. This is very different from the true figure of 40 volts.

This example illustrates an important point about any type of test equipment. It must be suitable for the job it has to do, and in most cases this means it must have the proper input impedance. This is especially true for AC, and the requirement gets more critical as the frequency rises.

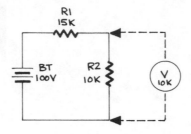

Figure T-1
Loading Effect of a Meter

Fortunately there are available today relatively inexpensive multimeters with very high input impedances, so this problem is more likely to arise when you are designing a piece of equipment that has a panel meter. Panel meters are more often *analog* than *digital*. In other words, they indicate by means of a pointer and a scale, not by a numerical display. The latter type is also available, but costs more.

The meter movement used almost universally for voltmeters, current meters, and ohmmeters is the *permanent-magnet moving-coil,* or d'Arsonval type, in which a small coil of wire is suspended between the poles of a permanent magnet. When a current flows through the coil the interaction between its magnetic field and that of the permanent magnet rotates the coil until the opposition of the hairsprings holding the coil balances the turning force. The pointer attached to the coil then indicates on the scale the strength of the current.

Since the wire used to wind the moving coil is very fine the meter by itself can measure only currents in the microampere or milliampere ranges. To measure higher values it must be provided with a *shunt,* which is a resistor connected in parallel with the movement that bypasses most of the current around it (see Figure T-2). The value R of this shunt is given by

$$R = R_m/(N - 1)$$

where N is the desired full-scale reading divided by the original full-scale reading (both in the same units), and R_m is the internal resistance of the meter.

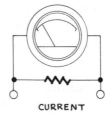

CURRENT

Figure T-2
Meter Shunt for Current Meter

In the case of a new meter R_m should be on the data sheet that comes with the meter, but if it is not, or if the meter is not new, the meter resistance can be obtained by using two rheostats and a battery. One rheostat and the battery are connected in series with the meter, as shown in Figure T-3, and the rheostat is adjusted until the meter reads full scale. The second rheostat is then connected in parallel with the meter, and adjusted until the meter reads half scale. The resistance of the second rheostat (required to make the meter read half scale) is equal to the meter resistance.

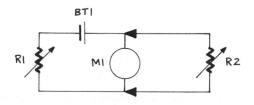

Figure T–3
Measuring Meter Resistance

A similar meter may be used as a voltmeter. In this case you have to connect a voltage-multiplier resistor in series with the meter so it will have the range you want (see Figure T-4). The value R of this resistor is given by

$$R = (V_{fs}/I_{fs}) - R_m$$

where V_{fs} is the full-scale reading required (in volts), I_{fs} is the full-scale current of the meter (in amperes), and R_m is the internal resistance of the meter (in ohms).

The loading effect of the voltmeter is given by

$$\text{Ohms per volt} = 1/I$$

where I is the full-scale current in amperes.

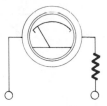

Figure T–4
Multiplier Resistor for Voltmeter

An AC voltmeter is obtained, using the same meter movement, by adding a rectifier diode in series with it, or connecting it in a diode bridge, as shown in Figure T-5. Deflection of

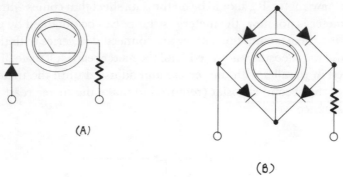

Figure T–5
Measuring AC [(A) and (B)]

the movement is always proportional to the *average value* of the rectified wave. However, meters are more usually calibrated to give the RMS value. The divisions on the meter scale are therefore made so that they are 1.11 times what they would be if the scale was calibrated for average value, but this is correct only where the AC is a sine wave.

To measure resistance a current must be passed through the resistor. For a fixed battery voltage this current will then be inversely proportional to the resistance value. The meter will show the current value in ohms. There are two ways of connecting the meter and battery: series and shunt. Both are shown in Figure T-6.

In the series type the current flowing in the circuit is given by

$$I = V/(R_x + R_1 + R_m)$$

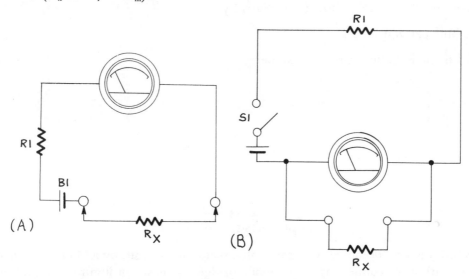

Figure T–6
Ohmmeters, Series and Shunt [(A) and (B)]

where V is the battery voltage, R_x is the resistance being measured, R_1 is the range resistor, and R_m is the internal resistance of the meter. If the input terminals are shorted together, R_x is zero, and I should be the full-scale current for the meter. This is why a series ohmmeter has the zero reading at the right-hand end of the scale.

The shunt-type ohmmeter has the zero at the left-hand end of the scale, because the resistance being measured is in parallel with or "shunts" the meter instead of being in series with it. When the input terminals are shorted together all the current goes through the short and none through the meter. The amount of current going through the meter is directly proportional to the value of the resistance being measured.

Other Test and Indicating Circuits

See the following list for other test and indicating circuits described elsewhere in the encyclopedia:

Absorption Frequency Meter
Ammeter Shunt
Amplified S-Meter
Battery Tester
Bridged S-Meter
Capacitor Tester
Demodulator Probe
Dip Meter
Electron-Ray Tube Tuning Indicator
Field Strength Meter
Hay Bridge
Kelvin Bridge
Marker Generator
Maxwell Bridge
Megohmmeter
Multimeter
Neon Indicator
Owen Bridge
Pilot Lamp
Resonance Bridge
RF Output Meter
Schering Bridge
S-Meter Addition
S-Meter Bridge
Transistor Tester
Vacuum-Tube Voltmeter
VSWR Meter
VU Meter
Wheatstone Bridge
Wien Bridge

Three-Channel Stereo Synthesizer
(Figure T-7)

Sometimes a third speaker is required between the standard left and right speakers. This can be provided by the use of a "Mix-n-Match" transformer (Alco Electronic Products, Co.), which has three windings of equal impedance (8 ohms). R1 and R2 are variable T-pads which are adjusted to obtain the desired blend of the left and right signals for the third speaker. See *Amplifier Circuits*.

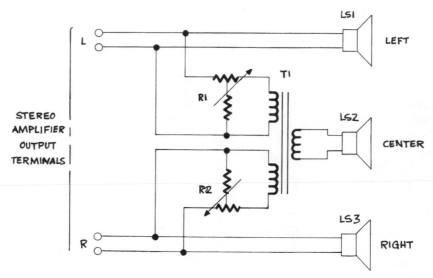

Figure T–7
Three-Channel Stereo Synthesizer

Three-Phase Rectifier (Figure T-8)

The two circuits shown here illustrate how rectification of a three-phase AC supply is done. In (A) the connections for a half-wave rectifier are shown. It is commonly used if DC output voltage requirements are relatively low and current requirements are moderately large. The DC output voltage is approximately equal to the phase voltage. However, each of the three arms must block the line-to-line voltage, which is approximately 2.5 times the phase voltage. For this reason it is desirable to use a three-phase half-wave connection only where one series unit per arm will provide the required DC output. The transformer design must be performed carefully, since there is a tendency for the core to saturate with unidirectional current flow in each winding.

The three-phase full-wave bridge rectifier circuit in (B) is often used if high DC power is required, and for better efficiency. The ripple component is 4.2 percent at a frequency 6 times that of the input frequency. The DC output voltage is approximately 25 percent higher than the phase voltage, and each arm must block only the phase voltage. See *Power Supply Circuits*.

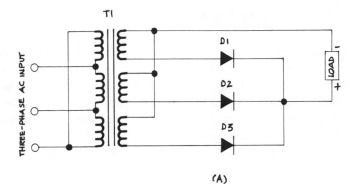

(A)

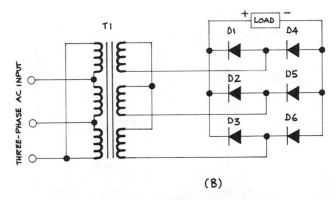

(B)

Figure T–8
Three-Phase Rectifiers

T-Network Filter (Figure T-9)

This is a low-pass filter. It may be used to match a transmitter output to the antenna. L1 is tuned to match the transmitter output impedance, and L2 is tuned to match the antenna impedance. See *Passive Filter Circuits*.

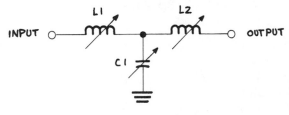

Figure T–9
T-Network Filter

Tone Control (Figure T-10)

This circuit provides both bass and treble tone controls. The treble control consists of R4 and C3. As the resistance of R4 is reduced it allows C3 to pass more and more of the high frequencies. The bass control consists of R2, C2, and C4. As the resistance of the part of R2 that shunts the larger capacitor is reduced, more of the signal bypasses this capacitor; at the time the resistance of the other part of R2 shunting the smaller capacitor is increased so that more of the signal goes through the smaller capacitor, and vice versa. The two capacitors present different values of reactance to the lower frequencies, so the setting of R2 determines the proportion of bass frequencies that are passed. These control circuits work best with a field-effect transistor (FET) because of its high input impedance. See *Amplifier Circuits*.

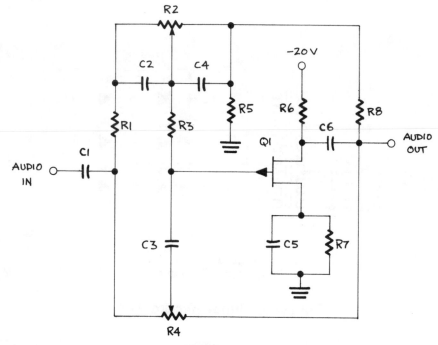

Figure T–10
Tone Control

Parts List

C1, C6	0.1 microfarad
C2	0.005 microfarad
C3	300 picofarads
C4	0.05 microfarad
C5	100 microfarads
Q1	2N3820 or equivalent

R1   100 kilohms
R2   1 megohm, variable (bass control)
R3   120 kilohms
R4   500 kilohms, variable (treble control)
R5   150 kilohms
R6   47 kilohms
R7   2.2 kilohms
R8   300 kilohms

Transducer Amplifier (Figure T-11)

A transducer samples another form of energy and converts it to electrical energy, so that the resultant signal is an electrical analogue of the original. Examples of transducers are strain gauges, thermocouples, and pressure gauges. Since a transducer's output very often is only a few millivolts, considerable amplification is required to increase the signal strength to the level where it can give a useful reading on an output device. It is also essential that the amplifier be extremely stable and free of distortion.

The example shown in the diagram uses an IC to amplify the signal from a magnetic transducer (pickup). See *Amplifier Circuits*.

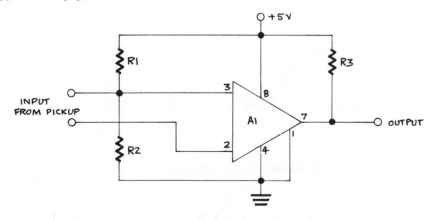

Figure T–11
Transducer Amplifier

Parts List

A1   Operational amplifier LM311 or equivalent
R1   4.5 kilohms
R2   1 kilohm
R3   2 kilohms

Transformer-Coupled Mixer-Balun
(Figure T-12)

This circuit is used to mix the unbalanced outputs of two microphone or phonograph-pickup preamplifiers, and feed the resulting signal to a balanced line. The secondaries of the transformers are connected in series, and the variable T-pads enable the signals to be blended in any manner desired. Since a T-pad presents a constant impedance to both the source and the load, its adjustment will not change the impedance match, nor will R1 exert any influence on R2, or vice versa.

The values of the variable resistors in the T-pads will depend upon the impedances involved. They are connected so that as one resistance value increases the other decreases, maintaining a constant impedance both ways. Suitable transformers for matching the source and load must also be selected. See *Attenuator and Pad Circuits.*

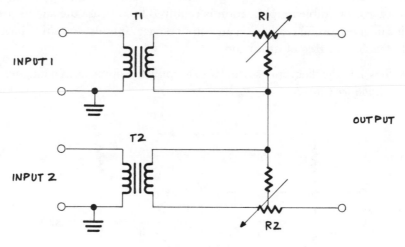

Figure T–12
Transformer-Coupled Mixer-Balun

Transformer-Coupled Push-Pull Power Amplifier
(Figure T-13)

This circuit was often used in audio applications, and could be biased for Class-A, Class-AB, or Class-B operation (see *Bias Circuits*). See *Amplifier Circuits.*

Parts List

T1   Interstage audio transformer with centertapped secondary and secondary-to-primary turns ratio of about 2 to 1

T2 Universal output transformer to match output load requirement of tubes to actual load (e.g., a loudspeaker)

V1, V2 6AR5 or equivalent

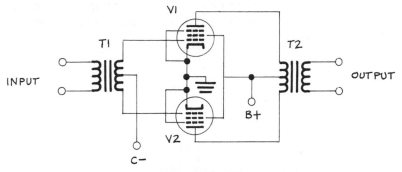

Figure T–13
Transformer-Coupled Push-Pull Power Amplifier

Transformer-Coupled Triode Amplifier
(Figure T-14)

This circuit is to all intents obsolete. In the early days of radio, when receiving sets were powered by batteries, it had the advantage that a lower plate voltage could be used, since no voltage had to be dropped across a plate resistor. Furthermore, people in those days were happy just to receive a radio program, without worrying about "hi-fi," which was still hidden in the mists of the future. Today, many people would find the distortive effects of interstage coupling transformers unacceptable. They are also more expensive than modern methods of interstage coupling and would add considerable weight to a receiver. See *Amplifier Circuits*.

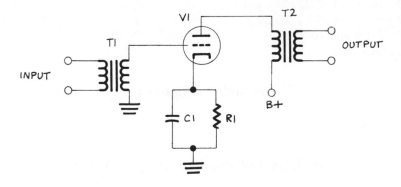

Figure T–14
Transformer-Coupled Triode Amplifier

Transistor-Amplified Crystal Detector
(Figure T-15)

The basic crystal set (q.v.) does not require a power supply, but improved reception is obtained by adding a transistor stage as shown in Figure T-15. A 3-volt battery supply is used to power Q1. See *Demodulation Circuits*.

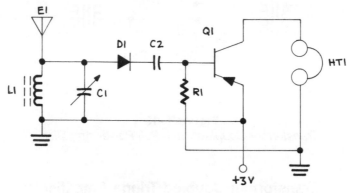

Figure T–15
Transistor-Amplified Crystal Detector

Parts List

C1	365 picofarads, variable
C2	0.02 microfarad
D1	1N34A or equivalent
E1	External antenna
HT1	Headphones, 2000 ohms minimum
L1	Internal loopstick antenna
Q1	GE84 or equivalent
R1	220 kilohms

Transistor Amplifier

See *Amplifier Circuits*.

Transistor AM Receiver (Figure T-16)

This circuit is similar to that of the transistor-amplified crystal detector (q.v.), but is more powerful. See *Demodulation Circuits*.

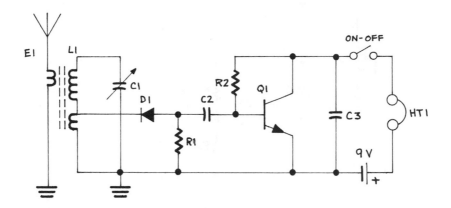

Figure T-16
Transistor AM Receiver

Parts List

C1	365 picofarads, variable
C2	0.02 microfarad
C3	0.01 microfarad
D1	1N34A or equivalent
E1	External antenna (if required) with ground and small extra coil on loopstick
HT1	Headphones, 2000 ohms minimum
L1	Loopstick antenna

Transistor Audio Oscillator (Figure T-17)

This useful little Colpitts test oscillator has a frequency output of 1 kilohertz, although it is possible to generate other frequencies by using different values for L1, C1, and C2. See *Oscillator and Signal Generator Circuits.*

Parts List

C1, C2	0.068 microfarad
C3	0.01 microfarad
L1	450 millihenries
Q1	2N2712 or equivalent
R1	100 kilohms
R2	120 kilohms
R3	2.2 kilohms
R4	100 kilohms

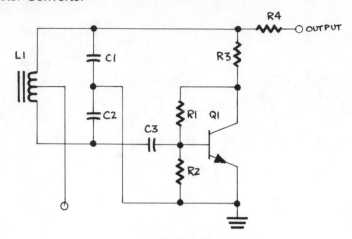

Figure T–17
Transistor Audio Oscillator

Transistor Converter

See *Converter*.

Transistor Dip Meter

See *Dip Meter*.

Transistor Ignition (Figure T-18)

This is an extremely simple circuit (many are much more complicated). S1 is the distributor points, T1 the coil.

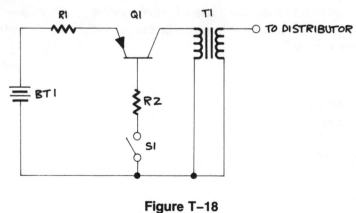

Figure T–18
Transistor Ignition

Parts List

Q1   GE34 or equivalent
R1   0.1 ohm, 2 watts
R2   270 ohms

Transistor IF Amplifier (Figure T-19)

A single NPN transistor is used in this IF amplifier. Its gain and band-pass are such that neutralization is not required. The input signal is coupled to its base through the IF transformer T1. To avoid loading down the resonant circuit by the low input impedance of Q1, the input signal is fed to a tap on the primary of T1, which is a low impedance point. The same is done with the output transformer T2.

R1 and R2 form a voltage divider to provide forward bias for Q1. C1, C2, and C3 are all RF bypass capacitors. See *Amplifier Circuits*.

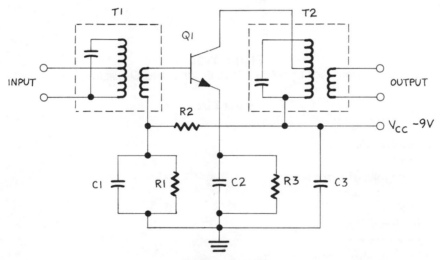

Figure T–19
Transistor IF Amplifier

Parts List

C1, C2, C3	0.05 microfarad
Q1	GE11 or equivalent
R1	390 ohms
R2, R3	150 ohms
T1, T2	IF stepdown transformer with tap on primary to match input impedance of following stage

Transistor Overtone Crystal-Controlled Oscillator
(Figure T-20)

This oscillator is used for aligning receivers. L1 may be constructed of 4 turns of #14 enameled copper wire, close wound on a half-inch diameter form. See *Oscillator and Signal Generator Circuits.*

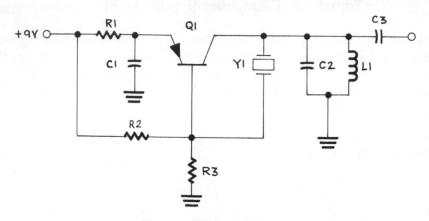

Figure T–20
Transistor Overtone Crystal-Controlled Oscillator

Parts List

C1	470 picofarads
C2	50 picofarads
C3	5 picofarads
L1	0.2 microhenries (see text)
Q1	T-1859 or equivalent
R1	330 ohms
R2	3.9 kilohms
R3	10 kilohms
Y1	48.98 megahertz, third overtone

Transistor Phase-Shift Oscillator

See *Phase-Shift Oscillator.*

Transmission Line Connections (Figure T-21)

Television and FM receivers have 300-ohm balanced inputs as a general rule. When the transmission line from the antenna is a twin lead (shielded or unshielded) the two wires are connected directly to the antenna terminals of the receiver, because an internal balun is provided to change the balanced signal path to an unbalanced one, as shown at (A). The balun in this example has a grounded centertap, but this is not always the case.

However, many people prefer to use a 75-ohm coaxial transmission line because it is shielded, which greatly reduces both noise pickup and losses to nearby metal objects, and is also more flexible than shielded twin lead. Usually "coax" requires the installation of a 300 to 75-ohm matching transformer both at the antenna connection and at the receiver input. Where the internal input transformer in the receiver has a grounded centertap, one matching transformer may be eliminated by connecting the coax as shown in (B), since the balun primary has an impedance of 75 ohms when only half of the winding is used. See *Antenna Circuits*.

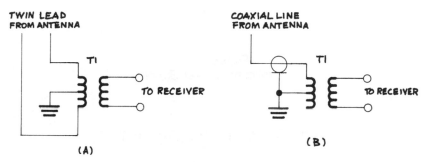

Figure T–21
Transmission Line Connections

Transmitter Power Supply (Figure T-22)

The more powerful a transmitter is, the greater the power it needs. Figure T-22 shows a power supply with an output of 1000 volts. As this is a full-wave supply using a centertapped secondary winding on the power transformer, the voltage across each half of the winding must be 1075 volts RMS, with a current capability of not less than 125 milliamperes. The filter chokes must be able to handle this current and be adequately insulated also, so that there is no possibility of the high voltage shorting between the winding and the core. Filter capacitors similarly must be able to withstand the high voltage. See *Power Supply Circuits*.

Parts List

C1, C2	2 microfarads, 1000 volts, paper
L1, L2	5/20-henry swinging choke
R1	20 kilohms, 75 watts
T1	High-voltage transformer, 1075 volts RMS each side, 125 milliamperes current rating
T2	Filament transformer, 2.5 volts, 5 amperes
V1, V2	866 rectifier tubes

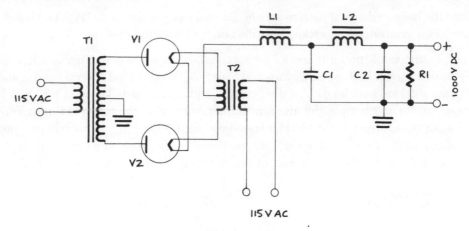

Figure T-22
Transmitter Power Supply

Treble Tone Control (Figure T-23)

In this circuit, when the sliding contact of R1 is at the extreme left it shorts out C2, and the larger value of C3 passes all signals at normal attenuation. However, when the sliding contact is at the extreme right it shorts C3, but now all signals have to pass through C2. C2's smaller capacitance offers an increasing reactance to lower signal frequencies, so that the higher frequencies are by comparison more accentuated. Intermediate settings of R1 select various degrees of treble response. See *Amplifier Circuits*.

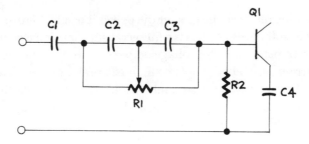

Figure T-23
Treble Tone Control

Parts List

C1   2.2 microfarads
C2   0.1 microfarad
C3   2.2 microfarads

C4   100 microfarads or circuit value
Q1   Audio transistor, according to circuit
R1   10 kilohms, variable
R2   500 kilohms or circuit value

Triac Switch (Figure T-24)

In this circuit a light duty switch S1 controls a triac capable of handling up to 40 amperes. When the switch is closed, a small current flows to the gate of the triac through R1. D1 then turns on, and current flows through the load R2. If no load is connected no current can flow to the gate, so the triac cannot be turned on.

The switch can be replaced by a transducer, such as a pressure switch, a thermal switch, a photocell, or a magnetic reed relay.

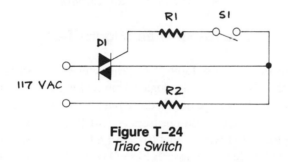

Figure T–24
Triac Switch

Parts List

D1   2N5444 or other suitable triac
R1   1 kilohm for 120 volts AC, 2 kilohms for 240 volts AC

Triac Switch with Gate Control (Figure T-25)

This circuit differs from the previous one by having a small transformer, such as a filament transformer, or a doorbell transformer, in the gate circuit. This allows the manual switch or transducer to be at a considerable distance from the triac. When the manual switch, which is across the secondary of T1, is closed, the reactance of the primary drops so that a gating current can flow to turn D1 on. The resistance of R2 must be chosen so that, in conjunction with the reactance of the transformer primary, the current flowing to the gate of D1 will be too low to trigger the triac when the secondary is not shorted.

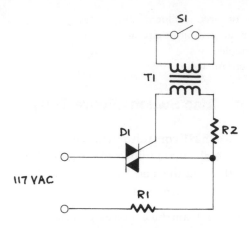

Figure T-25
Triac Switch with Gate Control

TTL NAND Gate (Figure T-26)

TTL stands for "transistor-transistor logic." The discovery that a single transistor could have more than one emitter gave rise to the TTL logic family. This gate is used in integrated circuits; therefore, the components shown are not discrete parts. However, they illustrate the way in which the gate works. If either or both emitters A and B are at ground potential (0), both the base and the collector of Q1 will be low (0), and a low bias will be placed on Q2. Q2 is therefore turned off, so its collector voltage is high (1), and its emitter low (0). If both inputs are high, as shown, the conditions are reversed, and Q2's collector voltage is low. This is a NAND situation. The gate is not limited to two emitters; as many as eight are used. See *Logic Circuits.*

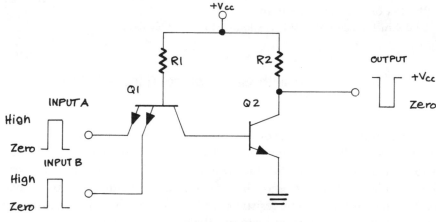

Figure T-26
TTL NAND Gate

Tuned-Plate, Tuned-Grid Oscillator (Figure T-27)

This oscillator gets its name from the fact that there is a resonant LC circuit at the grid input as well as at the plate output of the triode tube. When power is turned on electrons flow through the tube and into the side of C3 connected to the plate. At the same time, current begins to flow through L2. The resonant circuit C3-L2 begins oscillating, and at the correct moment a positive voltage pulse is fed back to the grid via the plate-grid capacitance.

The effect of this feedback voltage is to drive the grid in a positive direction, which increases current through the tube and charges C3 still further. However, the positive grid now attracts electrons from the stream of electrons passing from cathode to plate, and they start to flow back to the cathode via R1. This resistor has a high resistance, so at this stage electrons build up on the side of C2 connected to the grid faster than they can leak away through R1.

When the positive feedback signal is swinging the grid in a positive direction it is also coupled through C2 to C1, causing a positive charge to appear on the upper side of this capacitor. But as electrons build up on C2 the grid gradually becomes negative and the tube ceases to conduct. When this happens the feedback voltage disappears and C1 starts to discharge through L1. This causes the resonant circuit C1-L1 to start oscillating at a frequency determined by the inductance of L1 and the capacitance of C1.

Meanwhile C2 completes discharging through R1, and the grid loses its negative voltage, so that V1 can conduct again. The resonant circuit C3-L2 gets another current pulse to sustain oscillation, and in turn sends one back through V1 as before, so that the same cycle is repeated continuously as long as power is applied to the circuit.

C3 has to be adjusted so that the resonant frequency of C3-L2 is slightly lower than that of L1-C1 if the feedback voltage is to be in proper phase. The oscillations of the output circuit are coupled into the next stage via the resonant circuit L4 and C5.

This circuit is also seen with a crystal replacing L1, C1, and C2. See *Oscillator and Signal Generator Circuits*.

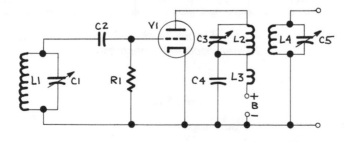

Figure T-27
Tuned-Plate, Tuned-Grid Oscillator

Tunnel Diode AM Transmitter (Figure T-28)

This CW (code) transmitter employs a tunnel-diode oscillator and a solar power supply. The frequency is determined by the crystal, and the L and C components are selected for resonance at the crystal frequency. See *Modulation Circuits*.

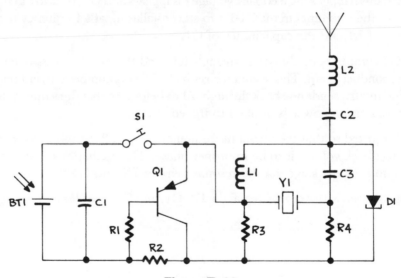

Figure T–28
Tunnel-Diode AM Transmitter

L1, L2   Selected for resonance at crystal frequency
Q1   2N404 or equivalent
R1   490 ohms
R2   22 ohms
R3   82 ohms
R4   220 ohms
S1   Telegraph key
Y1   Crystal for frequency of operation

Tunnel Diode FM Transmitter (Figure T-29)

In the absence of modulation this transmitter operates with a carrier frequency of 50 megahertz using the values given for the output tank. Modulation comes from the low-impedance microphone coupled via C1 to the base of Q1. The amplified audio signal is taken from Q1's emitter circuit and applied to the tunnel diode D1. Variations in the amplitude of the modulating signal vary the bias on D1, resulting in frequency deviations of the carrier signal. See *Modulation Circuits.*

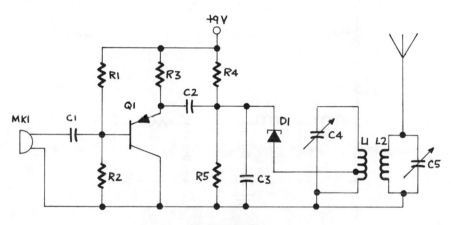

Figure T–29
Tunnel-Diode FM Transmitter

Parts List

C1, C2   10 microfarads
C3   800 picofarads
C4   70 picofarads, variable
C5   200 picofarads, variable
D1   Tunnel diode
E1   Rod antenna
L1   0.15 microhenry
L2   0.05 microhenry

MK1   Low-impedance (200 ohms) microphone
Q1   GE84 or equivalent
R1   220 ohms
R2   560 ohms
R3   82 ohms
R4   470 ohms
R5   10 ohms

Tunnel Diode Oscillator (Figure T-30)

In this circuit a varactor is used to adjust the frequency of the tunnel diode oscillator. The capacitance of the varactor varies in accordance with the control voltage, which has a range from 0 to 100. The oscillator frequency has a range of 12 to 22 megahertz. The bias for the tunnel diode is obtained from the 1.5-volt source, and this may be adjusted for best results by means of R4. See *Oscillator and Signal Generator Circuits*.

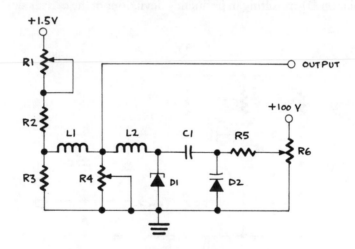

Figure T–30
Tunnel-Diode Oscillator

Parts List

C1   0.022 microfarad
D1   Tunnel diode
D2   Varactor
L1   1 millihenry (23 ohms DC resistance)
L2   1.5 microhenries (1 ohm DC resistance)
R1   250 ohms, variable
R2, R3   10 ohms

R4   100 ohms, variable
R5   10 kilohms, 1 watt
R6   10 kilohms, 1 watt, variable

TV IF Amplifier with IC (Figure T-31)

This circuit for the IF amplifier of a color TV uses an RCA CA3068 integrated circuit. The input from the tuner is coupled via the IF transformer T1 to the input at pin 6. There are several stages inside the IC. They are:

First picture IF cascode amplifier
Second picture IF amplifier
Third picture IF amplifier
Video detector
Video amplifier

The video output from the last stage emerges at pin 19. The interstage transformer T2 is connected between the output of the first stage (pin 9) and the input to the second stage (pin 13).

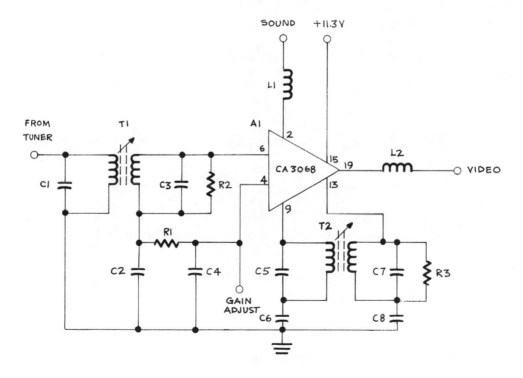

Figure T–31
TV IF Amplifier with IC

The IC also contains the sound IF amplifier, with its output at pin 2, and provides delayed AGC for the tuner at pin 7 (not shown). The power supply is connected to pin 15. The small chokes located in the sound and video outputs are self-resonating at the intermediate frequencies to prevent IF leakage into subsequent stages. In addition to the external components shown, there are, of course, the 47.25-megahertz and 41.25-megahertz traps required in all receivers. See *Amplifier Circuits*.

Parts List

A1	CA3068 PIX-IF System
C1	12 picofarads
C2	0.005 microfarad
C3, C7	18 picofarads
C4	10 microfarads
C5	8.2 picofarads
C6	0.001 microfarad
C8	0.01 microfarad
L1	12 microhenries
L2	10 microhenries
R1, R2	1.5 kilohms
R3	1.8 kilohms
T1, T2	Overcoupled IF transformers (center frequency approximately 44.0 megahertz)

UJT Relaxation Oscillator (Figure U-1)

This oscillator generates pulses with a repetition rate according to the time constant of R1 and C1. With the values given below, the rate is 1 kilohertz. The pulse width is determined by the base 2 inductance. With the value given below, the width is 12 microseconds. The rise and fall times are between 2 and 5 percent of the pulse width. See *Oscillator and Signal Generator Circuits*.

Parts List

C1	0.01 microfarad
L1	0.5 millihenry
Q1	2N1671 or equivalent
Q2	SE 7002 or equivalent
R1	100 kilohms
R2	470 ohms
R3	150 ohms
R4	270 ohms

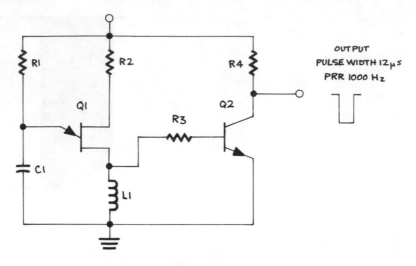

Figure U–1
UJT Relaxation Oscillator

Universal AC/DC Power Supply (Figure U-2)

When this earlier radio circuit was introduced there were still localities where power companies were delivering DC instead of AC. A power supply using DC cannot use a power transformer, of course, so the first requirement of a "universal" supply was that it could operate without one. In this example, as long as you had the power plug inserted so that the point A was positive, the diode V5 would conduct, and the power-line positive voltage would appear at B + + and B + when S1 was closed.

When used with an AC power line, V5 would conduct the positive excursions of the 60-hertz AC, but would not conduct the negative ones. The output from its cathode would therefore be a series of positive-going pulses, which the filter consisting of C1A, C1B, and R1 was required to smooth out into a passable resemblance to DC.

The five vacuum tubes used in these receivers had their heaters connected in series across the power line. The first figure, or figures, of tube numbers give the heater voltage required. Those shown in the diagram add up to 121 volts.

Five tubes, each requiring 24 volts, would add up to 120 volts. However, when this circuit was adopted, heater voltages had already been standardized at 6.3 or 12.6 volts, those being the battery voltages in use before the power line arrived. Consequently these tubes were used as much as possible. Only two new tubes were required, the diode with a heater taking 35 volts, and the audio output tube taking 50 volts. These higher heater voltages would have introduced hum into the signal circuits if they had been used in the converter, IF, and audio amplifier stages, but had no effect in the less sensitive stages.

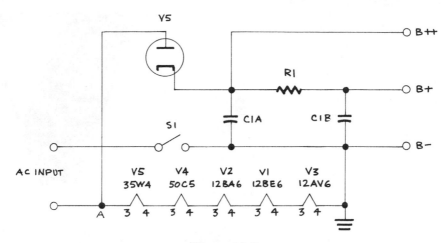

Figure U–2
Universal AC/DC Power Supply

The three 12-volt tubes draw their plate voltage from B + . This supply has negligible ripple. The audio output tube (50C5) gets its plate supply from B + + , which is unfiltered. The reason for this is that this tube draws a heavier plate current than the others, and it was desirable to minimize the current flowing through R1. Furthermore, since V4 was an audio power amplifier, fluctuations in the plate *voltage* would have very little effect (power amplifiers are *current* amplifiers).

Where higher plate current requirements arose, a choke had to be substituted for R1. This would have an inductance of about 5 henries. Another refinement was the connection of a pilot light across half of V5's heater filament. (Some tubes had tapped heater filaments so that they could be used on two different filament voltages.) This light would not only tell you if the set was on, but also would act as a fuse. The plate of V5 would be connected to the centertap so that if a short in the set caused a heavy current to be drawn the filament would blow and the pilot light would go out as well. See *Power Supply Circuits*.

Parts List

C1A, C1B	Two-section electrolytic capacitor, each section 40 microfarads, 150 volts
R1	1 kilohm, 2 watts

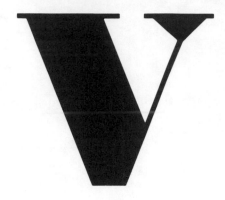

Vacuum-Tube Voltmeter (Figure V-1)

This multimeter used to reign supreme on the bench, but has been largely superseded by solid-state instruments. It consists of a bridge circuit with vacuum tubes in two of the arms. It is a very sensitive device, with negligible loading effect (its input impedance is on the order of 11 megohms).

The diagram shows a VTVM circuit for DC voltage measurement. The function of measuring AC voltage is added by converting the input AC to DC through rectification. Resistance measurement is also obtained by switching in a small battery with voltage so interconnected with the measuring terminals that the resultant DC voltage delivers correct ohm-current-voltage relationships across the grid of the input vacuum tube.

When no voltage is applied to the input both tubes have the same plate voltage, and no current flows through M1, which therefore reads zero. When a positive voltage is applied to the input, V2A conducts more heavily than V2B, since the latter's grid is tied to ground so that its bias is fixed. Consequently V2A's plate voltage falls but V2B's remains the same. This potential difference causes a current to flow through M1, and its pointer is deflected in proportion to the voltage being measured.

The instrument's probe contains a 1-megohm resistor R36. The arrow represents the probe tip. When it is placed in contact with the voltage being measured, current flows from the tip through R36 and the chain of resistors from R27 to R33, and back to the low side of the circuit via the common lead. The total resistance of this voltage divider is 11 megohms.

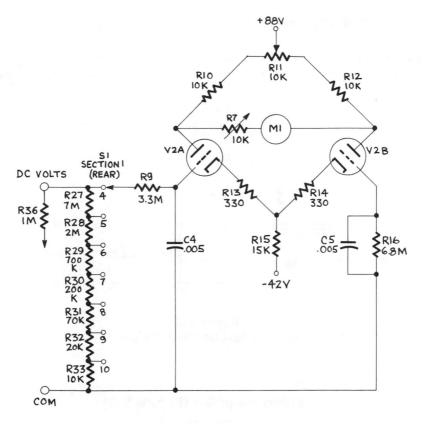

Figure V–1
Vacuum-Tube Voltmeter

These resistors are mounted on a rotary range switch, so that as the switch is turned the grid of V2A is connected via R9 in turn to each of the positions numbered 4 through 10. The operator selects the position that makes the voltage to be measured look to the meter like an input between 0 and 1.5 volts. The value indicated on the meter dial must then be multiplied by the range factor to get the correct value.

R11 is a front-panel control usually designated "zero adjust" for balancing the plate voltages of the tubes so that the meter reads zero when no measurement is being made. R7 is an internal calibration adjustment for the meter itself. The tubes used in this circuit are the two halves of a 12AU7A. See *Test and Indicating Circuits*.

Varactor Frequency Modulator

See *Frequency-Modulated Modulator*.

VHF Crystal-Controlled Oscillator (Figure V-2)

This crystal oscillator was designed specifically for overtone crystals, and will oscillate up to the eleventh overtone in the VHF range. See *Oscillator and Signal Generator Circuits*.

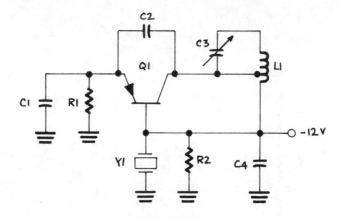

Figure V–2
VHF Crystal-Controlled Oscillator

Video Amplifier (Figure V-3)

This video amplifier circuit has an input impedance of 5 kilohms and an output impedance of 75 ohms. See *Amplifier Circuits*.

Parts List

C1	10 microfarads
C2	0.1 microfarad
C3	1000 microfarads
Q1	2N3704 or equivalent (with high cutoff frequency)
Q2	2N3702 or equivalent (with high cutoff frequency)
R1	10 kilohms
R2	1 kilohm
R3, R8	560 ohms
R4	100 ohms
R5, R7	47 ohms
R6	100 ohms, variable
R9	75 ohms

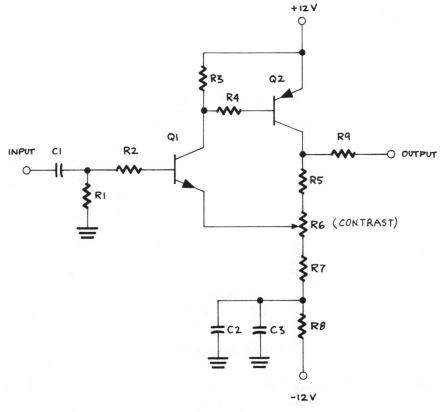

Figure V–3
Video Amplifier

Video Detector (Figure V-4)

This circuit is very similar to the second detector of an AM radio, with the exception of the added peaking coils L1 and L2. Their purpose is to compensate for the leakage to ground via distributed capacitance of the higher video signal frequencies. See *Demodulation Circuits*.

Parts List

C1   0.1 microfarad
C2   10 picofarads
C3   0.05 to 0.1 microfarad
D1   1N64

L1   36 microhenries
L2   250 microhenries
R1   1 megohm
R2   3.9 kilohms
T1   IF output transformer

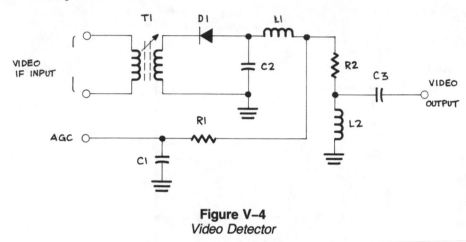

Figure V–4
Video Detector

Video Equalizers

See *Attenuator Equalizer; Phase Equalizer.*

Voltage Doubler (Figure V-5)

This circuit is sometimes referred to as the "conventional" voltage doubler to distinguish it from the *cascade voltage doubler* (q.v.). In this circuit, capacitors C1 and C2 are each charged during alternate half-cycles to the peak value of the alternating input voltage. The capacitors are discharged in series into the load, thus producing an output across the load of approximately twice the AC peak voltage. The ripple frequency is twice the supply frequency, and both capacitors are rated at the AC peak voltage. Filtering must be capacitor-input. See *Power Supply Circuits.*

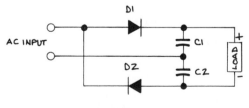

Figure V–5
Voltage Doubler

Voltage Regulator

See *Series Regulator*.

VSWR Meter (Figure V-6)

Voltage standing wave ratio (VSWR) is the ratio of standing wave maximum amplitude to standing wave minimum amplitude. Standing waves occur when part of the energy is reflected back along a transmission line from the load, due to an impedance mismatch. If there were no mismatch, the VSWR would be equal to 1. The increase in line loss is not too serious as long as the ratio is less than 2 to 1, but increases rapidly when it rises above 3 to 1.

In this circuit, the two diodes rectify the forward and reflected signals, which are applied to Q1 by the switch and, after amplification, to M1. R5 is set for minimum indication on the meter when the transmitter is turned off. Because R4 is in series with the transmitter output signal, this VSWR meter will cause an insertion loss and should be used only when checking the antenna system. See *Test and Indicating Circuits*.

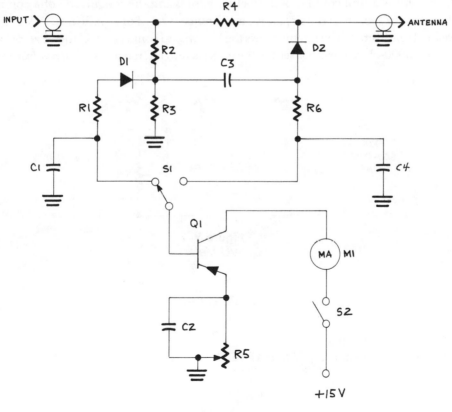

Figure V-6
VSWR Meter

Parts List

C1, C2, C3, C4	0.005 microfarad
D1, D2	1N34 or equivalent
M1	0-1 milliampere, DC or less
Q1	2N107, or equivalent
R1, R6	20 kilohms
R2, R3	47 ohms
R4	52 ohms
R5	5 kilohms, variable
S1	Double-pole, single-throw switch
S2	Single-pole, single-throw switch

VU Meter (Figure V-7)

A volume unit is a unit equal to a decibel for expressing the magnitude of a complex audio signal, as that of speech or music, above a reference level of one milliwatt. A VU meter scale is therefore logarithmic. The circuit shows the best way to connect a VU meter across a 600-ohm line, using a 3900-ohm T-pad for setting the meter's zero reference. See *Test and Indicating Circuits.*

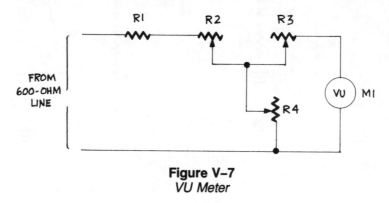

Figure V–7
VU Meter

Parts List

M1	VU meter
R1	3600 ohms
R2, R3, R4	3900-ohm T-pad (see *ATTENUATOR AND PAD CIRCUITS*)

Wavetraps

See *ANTENNA CIRCUITS.*

Wheatstone Bridge (Figure W-1)

This is a very sensitive DC bridge for measuring resistances between 0.1 ohm and 10 megohms. The A and B resistors are called the *ratio arms,* or *multiplier,* of the bridge. The variable resistor R_s is called the *rheostat arm,* or *variable standard,* of the bridge. The galvanometer G is frequently called a *null detector.*

The bridge is *balanced* when no current flows between points X and Y. This means the potentials at X and Y are the same. The null detector then shows a zero reading. To achieve this condition, the rheostat arm and the ratio arms are adjusted so that $A/B = R_x/R_s$. This gives

$$R_X = R_s A/B$$

R_X is equal to the resistance of the rheostat arm multiplied by the ratio A/B. The battery is usually 4.5 volts for general use, but it is important to keep in mind what the safe current is for the bridge so that the precision resistors used in it are not damaged. See *Test and Indicating Circuits.*

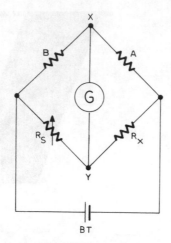

Figure W–1
Wheatstone Bridge

Wien Bridge (Figure W-2)

The Wien bridge is used for measuring frequency. It is similar to the Schering bridge (q.v.) and can be used to measure capacitance also. For measurement of frequency, if $C_X = C_S$, $R_X = R_S$, and $R_b = 2R_a$, then

$$f = (6.28C_SR_S)^{-1}$$

See *Test and Indicating Circuits*.

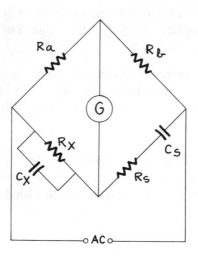

Figure W–2
Wien Bridge

Wien Bridge Oscillator (Figure W-3)

The principle of the Wien bridge (q.v.) is used in this audio oscillator, in which the six resonant circuits are Wien bridges. The maximum output of 100 milliwatts is adjustable by R8. See *Test and Indicating Circuits.*

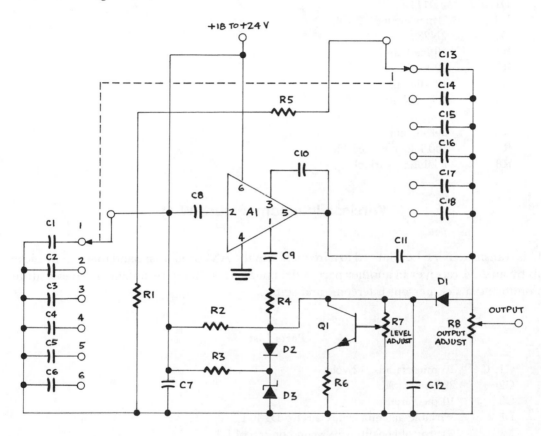

Figure W–3
Wien Bridge Oscillator

Parts List

A1	μA716C
C1, C18	0.33 microfarad
C2, C17	0.066 microfarad
C3, C16	0.033 microfarad
C4, C15	0.0033 microfarad
C5, C14	0.0022 microfarad
C6, C13	0.0016 microfarad

C7, C8, C9	10 microfarads, 25 volts
C10	100 picofarads
C11	15 microfarads, 25 volts
C12	0.1 microfarad
D1, D2	FD111
D3	Zener diode, 6.2 volts
Q1	2N2894
R1	10 kilohms
R2	20 kilohms
R3	12 kilohms
R4	4.1 kilohms
R5	5 kilohms
R6	2.4 kilohms
R7	100 kilohms, variable
R8	1 kilohm, variable

Wireless Intercom (Figure W-4)

This transmitter can be adjusted to a frequency in the AM broadcast band that can be picked up by an AM receiver in another part of the house. It is therefore suitable for baby-sitting, monitoring a sick person, intercom, and so on.

Parts List

C1, C4	10 microfarads, 12 volts
C2	220 picofarads
C3	10 picofarads
L1	Variable antenna coil, Calectro D1 841
L2	4 turns of hookup wire wound on top of L2
MK1	Crystal microphone
Q1	HEP 802
Q2, Q3	Calectro K4-507
R1	1 megohm
R2	4.7 kilohms
R3, R4	22 kilohms
R5	47 kilohms
R6	470 kilohms
S1	Single-pole, single-throw, on-off switch

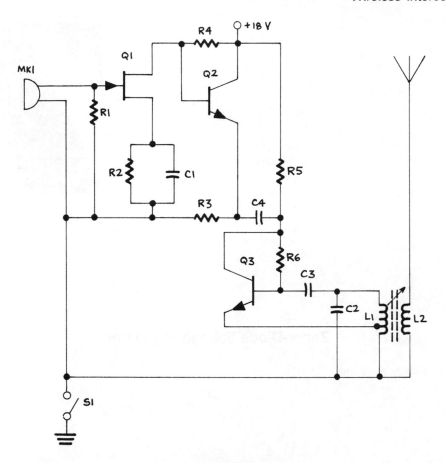

Figure W–4
Wireless Intercom

Zener-Diode Voltage Regulator

See *Series Voltage Regulator*.

PART II

Designing, Breadboarding, and Final Assembly

DESIGNING

Designing involves the arrangement and connection of parts to produce a complete electronic unit. It is often held to include the entire project, from the initial idea through to the finished article, but we shall use the term to cover the steps following the invention of a new circuit, or the utilization, with or without adaptation, of one already existing. In this sense, designing is the process of translating a schematic diagram into an assembly of hardware that performs the intended function of a device—a device that has until now existed only in the engineer's mind and in his notes.

Rather than discuss generalities, we shall take an actual circuit from Part I of this encyclopedia and proceed to fabricate it from the components given in its parts list. The circuit selected is that of the Laboratory Power Supply shown in Figure L-1. Many experiments can be performed using an adjustable power supply, and you will find this one, which you can make in an evening or two, very useful. Its parts list is repeated below in an expanded version that also cites Radio Shack catalog numbers. You can, of course, substitute items from any other supplier, or from your own collection accumulated from previous projects.

Shopping List

Quantity	Description	Catalog No.
1	741C operational amplifier	276–007
1	Wire-wrapping DIP socket, 8 pins	276–1988
2	Miniature electrolytic capacitors, axial lead, 3300 microfarads, 35 volts	272–1021
2	Miniature electrolytic capacitors, axial lead, 220 microfarads, 16 volts	272–1006
4	1N4001 silicon diodes	276–1101
1	1N4735 Zener diode, 6.2 volts	275–561
2	Nylon binding posts, red and black	274–662
1	0-1 mA DC panel meter	270–1752
1	0-15 V DC panel meter	270–1754
1	ECG249 Darlington pair	276–2042
1	Universal heat sink	276–1361
1	470-ohm resistor, 1/2 watt	271–019
1	3.3-kilohm resistor, 1/2 watt	271–028
1	10-kilohm linear taper potentiometer	271–1715
1	SPST potentiometer switch	271–1740
1	Power transformer, 120/12.6 V, 1.2 A	273–1505
1	Power transformer, 120/6.3 V, 0.3 A	273–1384
1	Perfboard, IC spacing, 2-3/4 × 6 inches	276–1395

A cabinet, control knob, decals, miscellaneous hardware, and so on, can be left for the present. You may have preferences of your own concerning the unit's external appearance, or you may want to duplicate the one described here. In either event, these details should be left until after the breadboarding stage.

About Component Parts

In selecting the component parts listed before, you have to be guided by various considerations. These were taken into account in making this list, but we'll take the time to review them now, so that you can tackle the other circuits in this book on your own. (Experienced readers may skip this section!)

The two most important and widely used components in almost any circuit are capacitors and resistors. The main use of a **capacitor** is to store electrical energy. It consists essentially of two plates of conducting material separated by an insulating material called a dielectric. If a capacitor is connected between the terminals of a battery, electrons flow on to the plate connected to the negative terminal, and are pulled off the plate connected to the positive terminal. When each capacitor plate is at the same potential as the battery terminal connected to it, electron flow ceases. The battery may now be disconnected, and the charges will remain on the plates; in this manner electricity is stored in the capacitor.

Although a direct current can charge a capacitor in this way, it cannot continue to flow once the capacitor is charged. A capacitor therefore can be used to block DC. However, an alternating current reverses the polarity on the plates each half-cycle, so it discharges them and then recharges them with the opposite polarity, and keeps on doing this as long as it is applied. Each time free electrons flow in or out of the plates according to the polarity of the current. Since this happens anyway in a wire carrying alternating current, the effect is as if the electrons actually flowed back and forth through the capacitor.

The amount of electricity that a capacitor can store depends upon its electrical size or capacitance. This is given in microfarads or picofarads. A picofarad (pF) is one-millionth of a microfarad (μF). The value is usually printed on the case. However, if the μF or pF is omitted, the size of the capacitor should tell you which it is.

Some of the larger capacitors are *electrolytic*. This means that their leads are *polarized*. The one marked positive (+) must go to a more positive connection point than the other (−) lead. They will fail if they are connected incorrectly. All capacitors have a voltage rating. This is the maximum voltage between its plates that a capacitor can stand without the dielectric breaking down. Again, it should not be exceeded.

Capacitors have many uses:

- They smooth rectified DC pulses into DC.
- They separate AC from DC.
- They are used with resistors to separate different frequencies, to perform timing functions, and to shape waveforms.

The most common type of **resistor** is called a composition resistor. It is made of carbon particles mixed with a binder, molded into a cylindrical shape, and hardened by baking. Wire leads are attached axially to each end, and it is then encapsulated in protective insulation. It will pass a current, but offers to it a degree of resistance corresponding to the ratio of binder to carbon particles.

The degree of resistance is denoted by three or four colored bands starting from one end of the resistor. The colors of the first three bands represent numbers from the following table:

COLOR	NUMERICAL VALUE
Black	0
Brown	1
Red	2
Orange	3
Yellow	4
Green	5
Blue	6
Violet	7
Gray	8
White	9

For instance, if the bands are brown, green, and red, its resistance value will be *1 5 0 0 ohms*. The first two bands give the first two figures and the third band gives the number of zeros following them (if this band is black, there are *no* added zeros).

If a gold band follows the three colored bands, the actual value of the resistor will be within 5 percent of the nominal value given by the colored bands. If a silver band is in this position, the actual value will be within 10 percent. No band means the actual value will be within 20 percent of the nominal.

It is worth mentioning in passing that in schematic diagrams the numerical value of a resistor is generally given without the word or symbol for ohms. It is assumed that this can be inferred from the graphic symbol itself. However, it is also common practice to add the letter K to the numerical value if the latter is in kilohms. (One kilohm equals 1000 ohms.)

Other types of resistor include those made by depositing a thin layer of resistive material such as a metallic oxide on a ceramic or glass core. The resistors used in microelectronic devices are deposited on the substrate by vacuum evaporation, vapor plating, or sputtering, using photomasks, screens, or stencils. However, many resistors used in IC's are passive transistors held to a certain conductance by the fixed bias on their bases.

Wirewound resistors form an important class. They are made by winding nickel-chromium alloy wire on a ceramic tube and covering it with a vitreous coating. They are able to withstand more heat than other types, but their spiral winding generates a magnetic field that limits their use to frequencies below one megahertz. This type has the value printed on the body of the resistor, which is considerably larger than the types previously described. For more information, see *Color Coding* in the *Appendix*.

Heat is produced in all resistors in proportion to the work done by the current in overcoming the resistance. Too much heat will damage or destroy a resistor, so it is given a power rating in watts, which tells the safe limits for current or voltage that may be applied to it, using the Ohm's Law formulas:

$$P = EI$$
$$P = I^2R$$
$$P = E^2/R$$

where: P = rate at which electrical energy is absorbed and converted to heat energy by the resistor, expressed in watts.

E = potential difference between opposite ends of the resistor, expressed in volts.

I = current flowing through the resistor, expressed in amperes.

If no power rating is specified you may assume that 1/4 or 1/2-watt resistors may be used.

The potential difference between the opposite ends of a resistor is also given by another Ohm's Law:

$$E = IR$$

where the terms have the same meanings as before. For instance, a resistor connected across a 10-volt power supply that has its negative terminal grounded, as in Figure II-1, will have a potential of 10 volts at one end and zero volts at the other. If the resistor has a resistance of 1000 ohms, you can determine the value of I by substituting these numerical values in the formula above:

$$10 = I \times 1000$$
$$I = 10/1000 = 0.01 \text{ ampere, or 10 milliamperes}$$

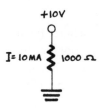

Figure II–1
Potential Drop Across Resistor

If this 1000-ohm resistor could be cut exactly in half and reconnected as in Figure II-2, the current would remain the same, since it would still be flowing through a **total** resistance of 1000 ohms. But each half-resistor would now be 500 ohms, so the voltage dropped across it would be:

$$E = 0.01 \times 500 = 5 \text{ volts}$$

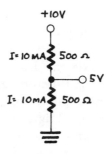

Figure II–2
Voltage Divider

This means that when you divided the resistor you also divided the voltage, so that its value at the point between the two half-resistors is 5 volts.

Suppose you had divided the 1000-ohm resistor so that one portion was 750 ohms and the other 250 ohms. The total resistance would still be the same, so the current would not change, but the voltage dropped across the larger portion would now be given by:

E = 0.01 × 750 = 7.5 volts

The voltage dropped across the smaller portion would be given by:

E = 0.01 × 250 = 2.5 volts

The result would be the same if you had used a 750-ohm resistor and a 250-ohm resistor that had not previously been part of a 1000-ohm resistor.

This is the way voltage dividers are made, choosing resistor values that will give the desired voltage at their junction. A voltage divider is commonly used to establish the bias on the base of a transistor, as in Figure II-3. (See also *Amplifier Circuits.*)

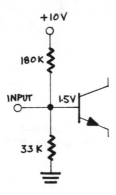

Figure II–3
Voltage Divider Used to Bias Transistor

Instead of fixed resistors you can use a potentiometer, as in Figure II-4. This is a resistor with a sliding contact that connects to any intermediate point on the resistor, according to how you set it. The ratio of the amount of resistance between the sliding contact and the upper end, and the sliding contact and the lower end determines the voltage at the sliding contact. A volume control uses this means to select the strength of signal to be applied to the audio amplifier. When it's turned all the way up, the sliding contact picks off the maximum signal voltage. As it is turned down the amount of signal voltage picked off gets less and less until when the sliding contact reaches bottom it is zero. Resistors are used in conjunction with capacitors in some important circuits known as *RC circuits.*

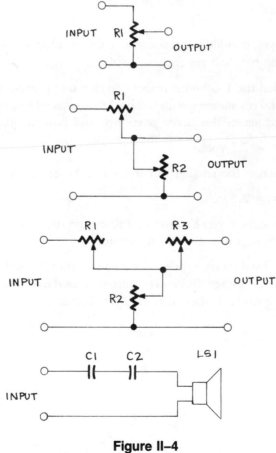

Figure II–4
Voltage Division by Potentiometer

Figure II-5 shows a capacitor, a resistor, and a switch connected in series with a battery. At a given instant we close the switch so that the voltage across the capacitor rises until it equals the voltage of the battery. As soon as the capacitor is charged to even a small voltage, this capacitor voltage opposes the battery voltage, reducing the charging current, which is

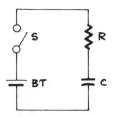

Figure II–5
Basic RC Circuit

proportional to the difference between the two voltages. The charging current therefore becomes less and less throughout the charging period, so that the capacitor voltage rises more and more slowly, as shown in Figure II-6. However, if the capacitor voltage had continued to rise at its starting rate it would have reached the final value at the point marked V '. Although the capacitor will have reached only 63.2 percent of the battery voltage at this point, it will have taken a certain length of time T called the **time constant** of the RC circuit of Figure II-5. This is given by:

$$T = RC$$

where: T   is the time contant, expressed in seconds,
             R   is the resistance expressed in ohms,
             C   is the capacitance expressed in farads.

        (In some cases it may be simpler to express T in microseconds, R in megohms, and C in picofarads, but be consistent.)

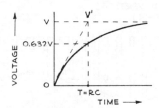

Figure II–6
Capacitor Charging Curve

This is a basic **timing circuit**, which charges in a length of time determined by R and C. The switch is often replaced with an electronic one, such as a transistor. An alternative circuit that charges the capacitor quickly and then discharges it slowly is shown in Figure II-7.

Other commonly-used RC circuits are integrators and differentiators, as in Figure II-8.

Many circuits employ **inductors.** An inductor is a spiral or coil of wire used to introduce inductance, the property that opposes any change in an existing current, into a circuit. Inductors may have iron or ferrite cores, or they may have air cores.

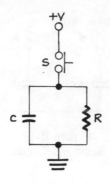

Figure II–7
Timing Circuit

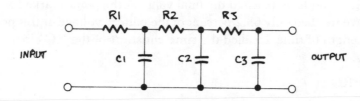

Figure II–8(A)
Differentiator

INPUT TV SYNC PULSES:
HORIZONTAL (NARROW), VERTICAL (BROAD)
OUTPUT INTEGRATED VERTICAL
TRIGGER PULSE (X—X)

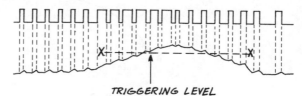

Figure II–8(B)
Integrator

The behavior of inductors is in many ways the opposite of capacitors. Like capacitors, they also store energy, but in the form of an electromagnetic field surrounding the coil instead of electrons on plates. This field disappears as soon as current flow ceases in the coil. Their opposition to the application of a current can be illustrated by a curve resembling that of the capacitor's voltage buildup in Figure II-6.

The most important circuits using capacitors and inductors are called **LC circuits.** These circuits resonate at frequencies that depend on the circuit values of inductance and capacitance. They are therefore important in oscillator and radio circuits. The resonant frequency of an LC circuit such as that in Figure II-9 can be determined from the inductance, capacitance, and reactance charts in the Appendix.

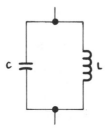

Figure II–9
LC Circuit

Inductors are also used in power supplies, mainly as line transformers, also as chokes, which pass DC readily but oppose AC. These uses are covered in the section on *Power Supplies* (q.v.).

Semiconductors are materials with properties that put them between conductors and insulators. Semiconductor devices (e.g., diodes and transistors) are now used extensively in all types of circuits and have replaced electron tubes except for certain specialized applications (e.g., picture tubes, high-power microwave equipment, TV cameras).

Diodes look something like resistors and are made by taking a single crystal of silicon or germanium and "doping" half of it with a substance such as arsenic, phosphorus, or antimony, and half of it with another substance such as boron, indium, or aluminum. This frees a large number of electrons that make it more conductive in the "forward" direction. This is the direction in which the junction between the two regions of the crystal is forward biased, but allows only a very small current to flow in the opposite direction ("reverse bias"). Diodes are marked so you can tell which is the forward and which is the reverse direction. One end, called the cathode, is indicated by a single band, or a diode symbol; or where several colored bands are used, by grouping them at the end. This end must have a voltage that is negative with respect to the other end if the diode is to conduct. Where colored bands are used the code is the same as that for resistors, except that all bands represent numbers in the type sequential number portion following the "1N" that is common to all diodes (and so not represented for that reason).

A diode will break down, or short, if too high a voltage is applied in the reverse direction. The maximum safe voltage rating is called the peak inverse voltage (PIV). Too high a voltage in the forward direction will result in too much current, which will burn the diode up. The maximum current rating must also be taken into account when selecting a type for a circuit.

However, the zener diode is a special type that breaks down when its characteristic PIV is exceeded, but recovers without harm to itself when the voltage drops below that level. Used in a circuit such as that in Figure II-10 it clamps the output at the zener voltage by bypassing the excess to ground.

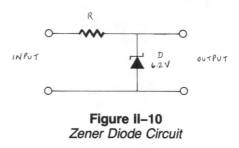

Figure II–10
Zener Diode Circuit

There are two principal types of **transistors**, bipolar and field effect. The bipolar has two types of materials similar to a diode, but has (usually) two junctions, and has three leads instead of two. Conduction between the emitter and the collector is controlled by the forward bias between the emitter and the base, which is from base to emitter with a PNP transistor, and from emitter to base with an NPN.

A field-effect transistor (FET) is unipolar. There are two main types of FET: the junction FET, or JFET; and the insulated-gate FET, or IGFET. The second-named is used in a host of ICs, and is alternatively named a MOSFET (metal-oxide-semiconductor FET). In the FET the electrodes corresponding to the emitter, base, and collector are called the source, gate, and drain. The input impedance to a bipolar transistor is relatively low (it is current operated), whereas that into a FET is very high, so in this respect it is like a triode vacuum tube (it is voltage operated). Transistors are used mostly in the common-emitter circuit, and a step-by-step procedure for designing this circuit is given under *Amplifier Circuits*.

Transistors are packaged in a variety of cases or "cans." The most popular kinds are shown in Figure II-11. Smaller ones, such as the TO-5, are used for general-purpose types that do not have to dissipate a great deal of power. Larger ones, such as the TO-3, are used for power transistors and are mostly mounted on heat sinks.

With the exception of the TO-3, all the transistors shown have three leads emanating in a straight line or in a semicircle, as viewed from beneath. The middle lead is always the base, while one of the other leads is distinguished by being placed nearer to the base lead or adjacent to some other feature, such as a small tab on the edge of the case, to denote that it is the emitter. The remaining lead must therefore be the collector.

The TO-3, however, seems to have only two leads. In this style, the collector is connected internally to the case. You can tell which of the two visible leads is the emitter by holding

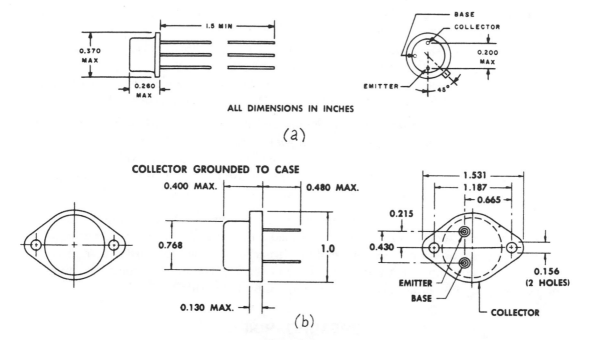

Figure II–11
Two Examples of Transistor Packaging
(a) TO–5 Small Signal Transistor
(b) TO–3 Power Transistor

the transistor with its underside toward you, and the two leads below the centerline (they are offset). The left-hand lead is then the emitter.

Integrated circuits have many transistors connected into microminiature circuits, all on a very small scale. If these transistors are MOS transistors they have extremely thin insulation (typically 0.1 micrometer thick) separating their gates from the substrate. On a dry day, ten steps on a nylon carpet can build up from 10 to 20 kilovolts of static electricity in your body. Simply getting up from a chair or stool with rubber casters can charge you with 10 kilovolts! After that, you need only point your finger at a MOS/CMOS IC to destroy it with a static discharge. Some ICs have gate protection circuits, but they can protect against only 1 or 2 kilovolts.

To be safe, not sorry, observe the following rules:

1. Don't store MOS ICs in nonconductive plastic bags, trays, foam, etc. Use conductive foam trays.

2. Place a MOS IC pin down on an aluminum foil sheet when you take one out of the conductive foam tray or out of a circuit.

3. Never remove any type of IC from a circuit when the power is on.

4. Use a battery-powered soldering iron to solder MOS ICs, not an AC-powered one. Better still, use IC sockets so that no soldering of the IC is required.

5. Do your work in a room with a tiled floor rather than a carpeted one, and if possible use a metal-topped table that you can ground. Ground you too!

6. Conditions become extremely hazardous when the relative humidity falls below 40 percent. If possible, refrain from handling MOS/CMOS ICs when it gets this bad.

Most ICs are enclosed in **dual-in-line packages,** usually abbreviated **DIPs.** These are plastic containers, about 3/4 inch long and 1/4 inch wide (for those with 14 or 16 pins). The pins are arranged in two rows coming out of the long sides, and bent downward for insertion into a socket or the holes in a perforated board. There are also DIPs with as few as 8 pins (single operational amplifiers) and as many as 40 or more (microprocessors, etc.).

When the DIP is the right way up (pins downward), the pin numbers run counterclockwise from pin 1 to the highest number. A mark called a pin locator is placed adjacent to pin 1. Alternatively, there is a notch in one end of the package. When the DIP is turned so that this is to your left, pin 1 is at the lower left-hand corner. The manufacturer's symbol and the device type number are also the right way up in this position.

BREADBOARDING

You are now ready to resume construction of the Laboratory Power Supply. When making a prototype of a newly designed device it is usual to **breadboard** it. This means making an experimental assembly to see if it works, while at the same time determining the best layout for the components and their connections. Since you want to use the same components in the final assembly, the prototype must be easily assembled and disassembled without damage to the parts, their leads, or their finish. The safest way to do this is to mount them on a perfboard and connect them by wire wrapping.

The perfboard in the shopping list is minimal size, so any piece that you have already should not be smaller. The prototype layout shown in Figure II-12 (a) does not include the front-panel items (meters, potentiometer with switch, and output terminals), nor the chassis-mounted item (Darlington circuit). You should install it on a heat sink until it is mounted permanently.

Mount all items loosely by inserting their leads through the holes in the perfboard. It will be necessary to bend those of the diodes and resistors. Do this carefully, so that the bends are not too sharp and the leads are not loosened in any way. The components should be resting on the board with the leads going through perpendicularly. Mount the transformers with machine screws and nuts (you'll have to drill larger holes for these).

This operation will be made easier if you have some kind of clamp to hold the board. (Radio Shack PC Board Holder, catalog no. 276–1568, is such an item.) This enables you to work on the board from both sides without having to hold it. You also will need a wire-wrapping tool, wire stripper, and wrapping wire. Wire from AWG #20 to AWG #30 is used and comes with insulation in various colors. We took advantage of this in building our power supply,

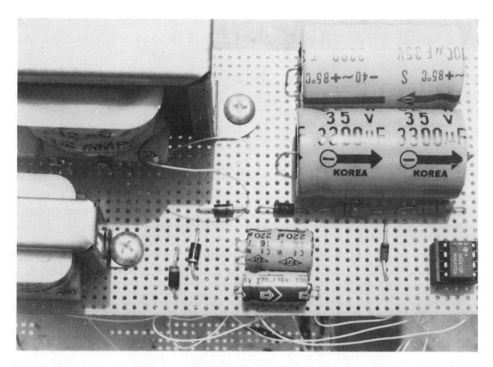

(a)

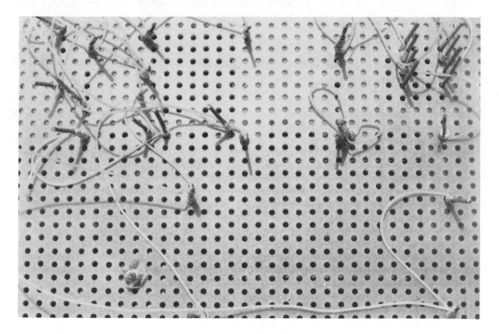

(b)

Figure II–12
Breadboard Layout

as it makes it easier to trace the connections, but you can't tell this from the photograph. You don't need to have more than one color, however, as long as the circuit is not unduly complicated. (See Figure II-12 (b).)

AWG #30 is popular for most circuits, but it is easily broken, so don't get too ambitious with it. Also, its current-carrying capacity is much less than that of ordinary hookup wire (AWG#20), so you may want to use the latter in a project where higher currents are involved. (The resistance of #30 is approximately 0.1 ohm per foot, compared to #20, which is approximately 0.01 ohm per foot.)

It is best at this point to trim off any excessively long leads, if you can see that the residual length will still be adequate, but you must leave enough projecting from the board to accept from five to seven wraps of wire around the pin or lead. This requires the removal of about 1 inch of insulation from the end of the wire.

When you have completed the assembly of the components on the perfboard, place it on a larger piece of plywood or strong cardboard, and place the other components on it, around the perfboard, and wire them.

At this time you should also install a current shunt across the terminals of M1, the 0-1 milliampere DC meter, so that it will be able to read 0-1 ampere. This shunt will divide the current between the shunt and the meter in the ratio of 999 to 1, so its resistance will have to be 1/999 times the meter resistance. When the power supply is putting out 1 ampere, 999 milliamperes will flow through the shunt and only 1 milliampere will flow through the meter. (Of course, if you have a 0-1 ampere meter, this will not be necessary.)

The best way to make this shunt is to take a length of the AWG #30 wrapping wire and wind it around a fixed resistor, soldering the ends to the resistor leads. The latter should have a resistance of at least 10 kilohms, so that its own shunting effect will be negligible. If your meter has an internal resistance of 50 ohms, the shunt will need a resistance of 50/999, or 0.05005 ohm. Since this wire has a resistance of 0.1052 ohms per foot, you can see that a length of 12(0.05005/0.1052) = 5.71 inches will be required. This length does not include the ends where they are soldered to the resistor leads.

If you don't know the internal resistance of the meter, don't try and measure it with an ohmmeter. The current from the latter will be more than sufficient to burn out the movement! Use the procedure given in Part I on *Test and Indicating Circuits*.

Finally, install the IC in the IC socket, observing all the precautions enumerated in the previous section. Go over the wiring carefully, making sure that all connections have been made correctly, and none missed. Using a yellow-colored pencil to draw a line along each wire on the schematic diagram as you check it not only ensures correctness, but eliminates the possibility of omitting a connection.

Before plugging the power cord into a service outlet, turn the potentiometer to its off position. Then, bearing in mind that there will be a potential of 117 volts present in the transformer primary circuits, place one hand in your pants pocket, and with the other turn the power

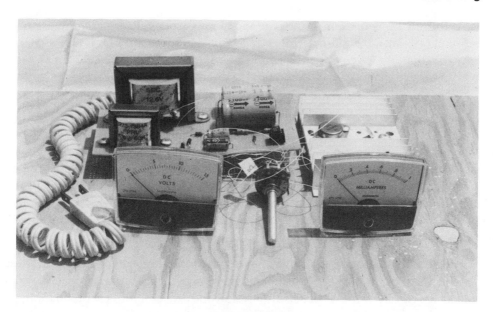

Figure II–13
Prototype Wiring

on. Watch for smoke! If there is any smoke or if you smell something burning or hear anything other than the click of the switch, suspect the worst and yank the plug. (See Figure II-13.) If nothing happens when you turn the power on, your assembly has at least passed the "smoke test." Now you should check the meters, which should both be reading zero. If the voltmeter reads 15 volts, or thereabouts, you have wired the potentiometer backwards. The milliammeter, now an ammmeter, should read nothing as long as nothing is connected across the output.

However, assuming that the voltmeter is reading zero, slowly rotate the potentiometer clockwise until the meter reads full scale. Then return it to the zero position. Now connect a short piece of wire across the output, and slowly rotate the potentiometer clockwise again, until the ammeter reads full scale. Don't rotate the potentiometer any further, but return it to the zero position. (The voltmeter should have been reading zero while you were doing this.)

Shut off the power and unplug the power cord. Feel all the components (warily!) to see if any have got unduly hot. The Darlington circuit in the TO-3 can may be warmish. This is permissible as long as it does not get hot. The potentiometer and diodes D1 and D2 should be cool. If the diodes are hot, check C1 and C2. If one of these is hot it is leaking badly and must be replaced. All other components should be cool.

The foregoing test should be repeated a few times, with various load resistors connected across the output. The lower resistances should be rated at not less than 25 watts. Using Ohm's Law, you can divide the voltage indicated on the voltmeter by the reading on the ammeter. This should always give you an answer approximating to the resistance. For example, if the voltmeter reads 15 volts, and the ammeter reads 0.25 ampere, the resistor must have a resistance of

60 ohms. With this same resistor, if the potentiometer is rotated counterclockwise until the voltmeter reads 10 volts, the ammeter must read 0.17 ampere, and so on. A few checks like this will verify that the circuit is performing as intended. You can now finish with the breadboarding stage, and proceed to the final assembly.

FINAL ASSEMBLY

There are several ways you can go. You can retain the wire-wrap assembly and house it in a suitable cabinet; you can disassemble it and rebuild it on a printed circuit board; or you can partially disassemble it, keeping part of the perfboard assembly and remounting the transformers on something more solid, such as a chassis. (If using a chassis, you could also build the whole thing on it, using hand wiring, and mounting the smaller components on terminal strips.)

The type of cabinet you decide to use will have some influence on your choice. These are usually either metal or plastic. The plastic ones may have metal or plastic covers or panels. The metal ones may have a metal chassis also. Metal ones are best for heat dissipation, plastic ones for general insulation. In cases where RF is present, metal cabinets give better shielding. In either case, the dimensions may be the critical feature. You will want a cabinet for your project that is large enough, but not excessively large.

Apart from the question of heat, metal cabinets are better where heavy components have to be mounted. A metal front panel is generally best for mounting controls, since it is thinner. A thick plastic panel may not allow enough threads to protrude for secure mounting of some components. For our finished power supply we selected an all-metal cabinet with an aluminum front and bottom, and decided to make a printed circuit board for the lighter components. (This was so we could cover the fabrication of PCBs, even though this circuit is quite simple.) The metal base would also make an excellent heat sink for the Darlington circuit, as well as a substantial platform for the transformers. An aluminum front panel is easy to drill and provides a good visual contrast for controls and their designations.

The colors of the cabinet influence the style of control knobs you choose. A black cabinet with an aluminum front panel, for instance, suggests black and chrome knobs, with black lettering for designations, and a black power cord.

Lettering on the front panel may be the hardest part of the whole thing. Very few people are sufficiently adept with a brush or pen to do this by hand. The most sophisticated results are obtained by silkscreening or decals. The latter are best for the amateur. Some people can do a passable job with embossed tape, but it lacks the professional look. Suitable decals can be obtained from amateur radio supply houses or from drafting supply stores. To prevent rubbing off, decal lettering should be protected by a covering of clear varnish.

Lettering on the front panel should be done after the holes have been drilled for the panel meters and controls. On this project there are two large holes for the meters, and four smaller mounting holes, for which templates are supplied with the meters. A centrally located hole

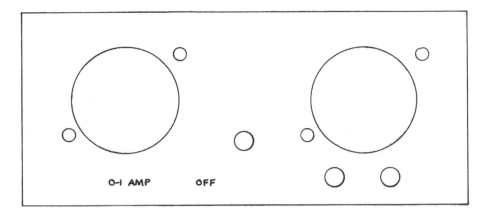

Figure II–14
Front Panel Layout

The two large (1.85" diameter) holes are for the meter bodies, and the four small (0.25" diameter) holes are for the meter mounting screws. The dimensions will vary with different meters. You should locate them according to the panel dimensions. Ours is 7.75" × 3.375", and our meter centers are 1.75" from the vertical panel edges and 1.5" from the top edge. The three holes for the potentiometer and terminals are 0.3125" (5/16") diameter.

has to be drilled for the potentiometer, and two smaller holes beneath it for the binding posts. The locations and dimensions of these for our panel are shown in Figure II-14.

After preparing the panel in this manner, apply the decals or whatever other means you have decided to use to designate the front-panel items. Place knobs, meters, and terminals temporarily in the positions they will occupy, so that you can locate the decals correctly. Then remove them and tape a piece of paper, plastic wrap, or other protective covering over the front panel to protect the artwork during the subsequent stages of assembly.

The main item in our procedure is the fabrication of the PCB. The easiest way to do this is to obtain a PCB kit and a direct-etching dry transfer. (If you get these from Radio Shack, the catalog numbers are 276–1576 and 276–1577, respectively.) However, we'll detail the procedure in a general way so that it will fit whatever materials you want to use.

Making a Printed Circuit Board. PCBs are available with copper on one side or on both. The first thing to decide is which to use. Generally, you will choose the single-side board because it is simpler to make and more reliable (most boards for use in outer space are single-sided). Double-sided boards are used where there is not enough room for the larger board area required when all components are on one side only.

Boards are made of paper-based and glass-based laminate. Paper-based boards are generally called **phenolic** and are by far the most widely used. They have excellent electrical qualities and relatively low cost, but they are unsuitable for use at temperatures over 105 degrees Celsius, or where moisture absorption would be a problem. **Epoxy-glass base** laminates are used where greater mechanical stress is involved. These are more moisture resistant and can withstand temperatures up to 170 degrees Celsius. You will find phenolic quite satisfactory for this project.

After settling the question of what type of board to use, and purchasing it, you are ready to begin. The first step is to design the layout of the board. For this, the most generally-used procedure amongst experimenters and hobbyists is to take a piece of paper the same size as the PCB and position all the components on it. This is so you can mark their terminal points. Components such as resistors that have pigtail leads must have their leads correctly bent first. If you used the perfboard procedure described under **breadboarding,** you will already have done this.

With a pin or sharp-pointed pencil, indent the terminal points so that you can see them plainly on the other side of the paper when you turn it over. This is because the circuit has to go on the opposite side of the board to the components. Now draw the connections between the terminals lightly with a pencil, avoiding crossovers. Where a connection is to a component that is not mounted on the board, such as a panel meter, the connection should be drawn to a convenient point to which a hookup wire can be connected.

When you are satisfied that you have all the connections sketched in correctly, you may place small adhesive circular disks over the terminal points (where holes will be drilled), and stick tape over the penciled connections. This will give you a professional-looking printed circuit template. However, the essential thing is to obtain a clear PCB design, like that in Figure II-15.

Prepare the copper surface of the PCB for etching. Some boards come with a transparent film over them that has to be peeled off. Others have a lacquer coating that must be removed

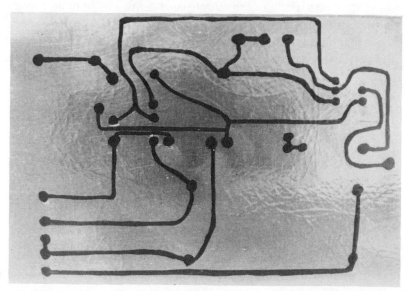

Figure II–15
Design for Printed-Circuit Board

by a *light* scouring with steel wool. The etchant cannot dissolve all the copper to be removed if the surface is not clean and bright.

Now get a piece of carbon paper the same size as your PCB. Tape it and the design to the board so they won't slip, and trace the circuit on to the copper surface with a ballpoint pen. Check the diagram to make sure all lines have been drawn. Then use a center punch to indent the copper surface where the holes are to be drilled.

Apply the **resist** to the board. Resist is a fluid that, when dry, will prevent the etchant from removing the copper from areas covered by it. The simplest way of applying it is by means of a pen (obtainable from Radio Shack or other supplier, either separately or as part of a PCB kit). Make sure the ''ink'' is emerging from the pen (it makes a black line), then trace over the lines on the copper surface, making heavy dots where the holes are to be drilled. When the resist has dried (about a minute), go over the lines and dots again to ensure complete coverage.

The **etchant** is a solution of ferric chloride, which is an orange-colored liquid that can stain your skin (and anything else it comes in contact with). It is also poisonous and corrosive, so you shouldn't drink it or breathe the fumes. It would be wise to wear rubber gloves and an apron, and perform the etching in a well-ventilated area.

You need a developing dish a little larger than the PCB. Put the PCB in it, and cover it to a depth of 1/4 inch. Agitate the liquid back and forth, as you would in developing a photographic print, until the copper is all etched away from the parts of the board not protected with resist. This will take 15 to 20 minutes. When all the unwanted copper has been dissolved, remove the board with tongs, pliers, or rubber gloves, and hold it under the cold water faucet for about two minutes. This will wash off the etchant and stop the chemical action. Don't keep the used etchant; it isn't any good for further etching.

The next step is to remove the resist, which is done with **resist ink solvent.** This is toluene, which is even worse than the etchant, since it also flammable. Don't use the same dish you used for etching. Remove the final traces by scrubbing with household cleanser or fine steel wool under running water until the copper is bright and shiny.

Your PCB is now ready for mounting the components. The first step is to drill holes on the dots so that the pigtail leads can be inserted. Use a 1/16-inch drill bit. Then drill a hole in each of the four corners of the board for mounting it. Use a 3/16-inch drill bit. The components are mounted on the opposite side of the board from the copper circuit. The leads come through the holes and are soldered to the copper conductors, using a small iron (excess heat will cause the copper to lift from the board). Make sure the tip of your iron is firmly seated in the heating element, and keep it clean by frequent wiping. Use the least amount of solder necessary, but be sure it flows and ''wets'' the lead and copper conductor properly. Also solder into their holes the leads required to connect the board to the components that are mounted on the panel or the chassis.

After you've finished the soldering, inspect the board for any solder ''bridging'' between conductors, and any damaged circuits. If you've performed the previous step carefully, there should be none. Cut the pigtails short, remove any excess flux and clean the board with some more of the resist ink solvent, as before. As a final step, spray both sides of the board with a protective coating of **polyurethane resin.** This will prevent oxidation of the copper conductors.

The PCB should be mounted component side uppermost on the chassis (or base plate) with four standoffs inserted through the four corner holes you drilled for the purpose, and four corresponding holes drilled in the chassis. The standoffs should raise the board about an inch clear of the chassis.

Figure II–16
Completed Project (Interior)

Figure II–17
*Completed Project
(Exterior)*

PART III

Performance Testing and Troubleshooting

PERFORMANCE TESTING

Performance testing means checking that a device is capable of fulfilling the purpose for which it was designed. The details are called its specifications. If the check is satisfactory, there is no need to do more. If it is not, then troubleshooting is required.

Some devices require very little in the way of checking. A light bulb need only be turned on to see if it lights. But others must be compared with test instruments to verify that their output voltage, current, frequency, or other parameter is within the limits specified. For instance, the laboratory power supply of Part II is supposed to have a maximum output of 15 volts DC with a maximum current of 1 ampere. It is also supposed to be able to maintain this output despite variations of load or line voltage. Checking this calls for:

(1) Checking the readings of the panel meters against test instruments of known accuracy;

(2) Applying a variable load to the output to verify that the voltage set by the control potentiometer remains constant in spite of load variations;

(3) Varying the input from the power line to check that the output voltage remains constant while this is being done.

For this you would need:

(1) A voltmeter, preferably of greater accuracy than the panel voltmeter;

(2) A current meter, preferably of greater accuracy than the panel ammeter;

(3) A variable resistive load capable of absorbing the maximum power output of the supply (15 watts);

(4) A means of varying the line voltage between 105 and 125 volts AC.

The procedure for checking the panel meters is as follows:

(1) Place the laboratory power supply on the bench in its normal operating attitude.

(2) Before applying power to the power supply, check that each meter's pointer is resting directly over the zero mark. If not, reset it, using the screwdriver adjustment on the front of the instrument. It is preferable to turn this so that the pointer is first shifted upscale, and then brought downscale to the zero mark. Tap the glass *lightly* while doing this to eliminate friction error.

(3) Now connect a DC voltmeter to the output terminals. Since the panel meters have an accuracy of ± 5 percent, it would be best to use an instrument with an accuracy of ± 0.5 percent or less. Then you have to take into consideration only the error of the panel meter. Since this is given as ± 5 percent of *full scale,* it means that the reading will not be off by more than 5 percent of the full-scale value of 15 volts, or ± 0.75 volts. In other words, a reading between 14.25 volts and 15.75 volts is acceptable when the voltage is actually 15 volts. When the voltage is 10 volts, the acceptable reading is between 9.25 and 10.75 volts, and so on.

If your test voltmeter is not ten times more accurate than the panel meter, you have to allow for its error also. This means subtracting its tolerance from that of the panel meter and using the lesser percentage. For instance, if your test voltmeter has an accuracy of ± 3 percent, then the tolerance you have to use is $\pm 5 - \pm 3 = \pm 2$ percent. Acceptable readings could then only be between 14.7 volts and 15.3 volts.

(4) Verify that the voltage at the AC outlet is between 110 and 120 volts. Plug in the power supply and turn on the power.

(5) Rotate the control knob clockwise and set the output voltage to the following values, as read on the panel voltmeter. The test voltmeter (if ± 0.5 percent) shall read within the limits shown:

Panel Meter (volts)	Test Voltmeter (volts)
5	4.25 – 5.75
10	9.25 – 10.75
15	14.25 – 15.75

(6) Rotate the control potentiometer knob completely counterclockwise and turn off the power.

(7) Disconnect the test voltmeter and replace it with the test ammeter. Connect a test lead between the output terminals.

(8) Turn on the power and rotate the control potentiometer knob clockwise. Set the output current to the following values, as read on the panel ammeter. The test ammeter (if ± 0.5 percent) shall read within the following limits:

Panel Meter (amperes)	Test Ammeter (amperes)
0.25	0.20 – 0.30
0.50	0.45 – 0.55
0.75	0.70 – 0.80
1.00	0.95 – 1.05

(9) Rotate the control potentiometer completely counterclockwise and turn off the power.

(10) Connect a 100-ohm, 15-watt, rheostat to the output terminals, setting the rheostat for maximum resistance. If a rheostat is not available, connect the following resistors in turn to the output terminals, as shown in the following table: 75 ohms, 3 watts; 50 ohms, 5 watts; 25 ohms, 10 watts; 15 ohms, 15 watts.

(11) Turn on the power.

(12) Rotate the control potentiometer clockwise until the panel voltmeter reads 15 volts.

(13) Rotate the rheostat to reduce resistance for the following readings on the panel ammeter. The panel voltmeter reading shall not change by more than ± 0.25 volt during this test:

Panel Ammeter	Equivalent Fixed Resistor
0.2 ampere	75 ohms
0.3 ampere	50 ohms
0.6 ampere	25 ohms
1.0 ampere	15 ohms

(14) Leaving the rheostat connected, turn off the power.

(15) Unplug the power cord and plug in a Variac, Powerstat, or similar line voltage adjustment autotransformer in its place. Plug the power cord into the autotransformer and adjust the latter for 115 volts.

(16) Turn on the power and check that the panel meters are reading 15 volts and 1 ampere. If they are not, reset them with the control potentiometer.

(17) Adjust the autotransformer until the input reads 105 volts AC. The panel-meter reading shall not change by more than ± 0.25 volt.

(18) Adjust the autotransformer until the input reads 125 volts AC. The panel meter reading shall not change by more than ± 0.25 volt.

(19) Turn off the power and disconnect the test equipment.

If you don't have a Variac, or equivalent, consider making the **line voltage adjuster** described in Part I. With a 12.6-volt centertapped secondary you can raise or lower the input voltage by 6.3 volts. While this is not as good a test as that described above, it will save buying an expensive autotransformer.

The foregoing procedure is a typical performance check. A similar procedure would be used for periodic calibration for equipment with internal adjustments. First, a performance check, then adjustment, where necessary, to get the performance within specification limits, and then another check to confirm that the adjustments have been done correctly.

As mentioned already, test instruments preferably should have at least ten times the accuracy of the device being tested. However, most experimenters don't need this much accuracy, and will be quite content to know that their project works reasonably well. For most people the following list of inexpensive test equipment is sufficient:

Multimeter
Signal injector
Laboratory power supply (home brew)
Logic probe
Transistor checker

Three other items which are nice to have, but cost a lot more, are an oscilloscope, a counter, and a signal generator. You will find these, and some others, more essential for a TV, CB, or amateur radio.

TROUBLESHOOTING

There are fundamental differences between troubleshooting integrated-circuit equipment and equipment employing discrete components. Discrete components have simple characteristics, such as resistance, capacitance, and inductance, that are easily checked by traditional troubleshooting tools. But when the circuit is microminiaturized and encapsulated, individual components are no longer accessible. The only thing you can test is the performance of the entire circuit.

The classical method of isolating the cause of a failure consists of four steps:

1. Determine symptoms
2. Determine section
3. Determine circuit
4. Determine component

Since you can't do Step 4, it seems your task has become simpler. This is also true to some extent of linear ICs whose performance can still be evaluated by use of signal generator and oscilloscope. But since the majority of ICs are digital, with numerous inputs and outputs to be stimulated and monitored simultaneously, Step 3 can become extremely time-consuming. If you have to disconnect leads and unsolder pins as well, the resultant wear and tear is likely to have further adverse effects on boards and printed circuits, and increase the chances of static burnout.

Before getting too ambitious, there is one check that should always be made. Verify that all ICs are seated properly in their sockets, if they are so installed, and that their pins are not corroded or in some way making poor contact. This also applies to board edge connectors, cable connectors, and so on. A great deal of trouble in PCBs with ICs is traceable to connections, which is why equipment that has to be highly reliable uses gold-plated pins.

Troubleshooting Digital ICs

Linear circuits handle analog signals with voltage profiles corresponding to values that vary continuously. This is why the oscilloscope is the preferred instrument for studying them. Digital circuits, on the other hand, operate by switching between states—from high to low, and vice versa. To be in the high state is to have a potential above the high threshold; to be in the low state is to have a potential below the low threshold. The *exact* value of the potentials is unimportant as long as they are above or below the respective thresholds, since this is all that is needed to actuate the circuit. If a sufficient stimulus is applied to the input, the circuit will produce the correct output every time if it is working properly. (Potentials *excessively* above or below the respective thresholds may damage the circuit.)

From this it becomes clear that the essential requirement for troubleshooting equipment with digital ICs is a means of verifying the presence of the correct voltage states on the various pins of the different ICs, without trying to remove them from the board. This not only saves time and minimizes the risk of damage, but also allows us to examine the IC's performance in its natural environment. First, though, let's consider types of failures to look for and the symptoms produced by them.

Common Defects

IC failures fall into two main categories: external and internal. External failures occur in the connections between ICs, or in discrete components connected to them. These are repairable. Internal failures are not.

The most common internal defect is an open bond between a pin and the chip. When this happens to an input bond, as in Figure III-1, the correct signal will be present on the pin, but cannot reach the chip. The input of the chip will float to a "bad" level between the high and low thresholds, which a TTL gate will see as a permanent high state. The effect on the output will depend upon the nature of the circuit. If we take as an example a NAND gate with two inputs, with a normal truth table as at (a), the effect of an open bond at A will be as in the truth table at (b).

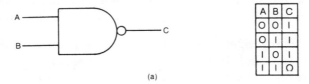

A	B	C
0	0	1
0	1	1
1	0	1
1	1	0

(a)

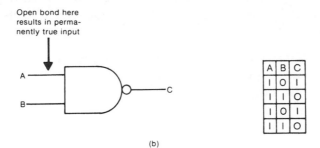

Open bond here results in permanently true input

A	B	C
1	0	1
1	1	0
1	0	1
1	1	0

(b)

Figure III–1
Effect of Open Bond

An open output bond will block the correct output from the chip, so that it will not appear on the output pin. This will also go to a "bad" level, and will affect any and all input pins on other ICs to which it may be connected. These will behave as though they had open input leads, although the "bad" level will now be present on the input pin. This would lead you to look further back for the origin of the malfunction.

Instead of opening up, the input or output bond may short to V_{CC} or ground. If the short is to V_{CC}, all signal lines connected to that point will be high. Conversely, a short to ground will hold them to a low state. The most likely effect of either condition is to inhibit all signals normally to be found beyond this point, so it is one of the easiest troubles to track down.

It is not so easy to analyze a problem caused by a short between two pins when neither is V_{CC} nor ground. When either pin goes to a low state it pulls the other with it, yet when the two pins should be high or low together they show the proper voltage.

An internal failure in the chip is always catastrophic to its performance, so that the output pins are locked high or low and will not change in response to appropriate stimuli. This is because the failure blocks the signal flow by completely preventing switching action.

An open signal path in the external circuit has the same effect upon the input to which it is connected as an open output bond in the previous IC. The input will float to a "bad" level. However, since the correct signal appears on the output pin of the previous IC, we know that the interruption must exist between that output pin and the input pin of the following IC.

A short between an external signal path and V_{CC} or ground exhibits the same symptoms as an internal short of the same kind. If the short is to V_{CC} the signal path will be high at all times; if to ground it will be permanently low. This one is hard to isolate, and only a very close inspection of the circuit will determine whether the fault is internal or external.

Test Equipment

The *digital logic probe* is an expensive but highly useful piece of equipment. It has two leads with alligator clips that connect to the circuit's V_{CC} and ground. When the probe tip is touched to one of the pins of an IC, or to some other point in the circuit, light-emitting diodes on the probe indicate the high, low, or pulsed (to 10 megahertz) logic states present on that pin.

You can also use it to test an IC by applying the proper voltage to an input pin and observing what happens at the output. (You could do this almost as well with a voltmeter.) However, it works even better with a *logic pulser,* which looks like a probe, but which is used to inject a narrow pulse into the circuit wherever you apply it, so you can observe on your logic probe what happens at any output.

Another useful device is the *logic clip*. This resembles a broad clothespin. It is made to clip over a 14- or 16-pin DIP in such a way as to contact all of its pins simultaneously. It displays the condition of all the pins at the same time by means of LEDs that light for each pin above a threshold level.

A somewhat different instrument is the *logic comparator*. Like the logic clip, this device is clipped over a DIP so as to contact all of its pins at once, but it is made so that you can install in it another (good) DIP of the same type as that being checked. The reference IC therefore gets the same signals and voltages as the circuit IC. If there is a discrepancy, the different pin is indicated by a LED. If everything is correct, no light appears.

For more elaborate equipment, such as that containing microprocessors and large numbers of other DIPs, the test items just described are quite inadequate. The job is too complex. For troubleshooting on this scale you would use a *logic-state analyzer*. This instrument displays the actual data being processed, "freezing" it to allow examination of the data words or word containing the anomaly. Unless you are building something comparable to a computer you would have no use for a logic-state analyzer, which is also very expensive.

Troubleshooting Linear Equipment

Linear circuits handle voltages that vary continuously instead of switching between high and low states. In a great many cases the shape of the waveform or the frequency response is very important; hence, signal generators and oscilloscopes are often used. One good thing about linear electronic equipment is that all the circuits of any given type work on the same general principles, can suffer the same defects, and are amenable to the same troubleshooting techniques, whether they are in the form of ICs or made up of discrete components. In fact, if we consider troubles in broad rather than limited terms, specific defects may cause the same problems in all types of equipment.

The *physical manifestation* of the defect may vary considerably from one type of equipment to another. For instance, an IC used as an amplifier can cause distortion due to a "leaky" transistor. This broad category (distortion) shows up in different ways according to what the amplifier is being used for:

1. In the video section of a TV set the distortion may show up as a *smeared picture*.
2. In an audio amplifier or radio receiver it may show up as *garbled sound*.
3. In a radio transmitter it may show up as *excessive harmonic radiation*.
4. In a test instrument it may show up as *inaccurate readings*.
5. In an item of industrial control equipment it may show up as *erratic operation*.

Figure III-2 applies to all types of electronic equipment. It is used to select a suitable diagnostic test technique. To use this chart, first determine the general nature of the problem. This must be deduced from its broad technical characteristics. For instance, if the equipment *doesn't work at all* it is DEAD, whether the item is a receiver, audio amplifier, industrial control, transmitter, or test instrument. If the equipment *works, but lacks its usual response,* it is WEAK. In a radio transmitter this may show up as lowered power output, in a receiver as lack of sensitivity, in control equipment as sluggish operation. If the equipment *works "now and then"* or if some other malfunction "comes and goes," it is considered INTERMITTENT. If the equipment *works, but not quite properly,* the complaint is DISTORTION, as in the examples given above.

UNIVERSAL TROUBLESHOOTING CHART

DIAGNOSTIC TECHNIQUE / COMPLAINT	VISUAL INSPECTION	CHECK POWER SUPPLY	VOLTAGE CHECKS	SIGNAL TRACING	SIGNAL INJECTION	BRUTE FORCE	PARTS SUBSTITUTION	COMPONENT TESTS	ALIGNMENT	REMARKS
DEAD	X	X	X	X	X			X		
WEAK	X	X	X	X	X				X	Align only if other tests indicate
INTERMITTENT	X	X				X	X	X		
DISTORTION	X	X	X	X				X		"Weak" and "distorted" may be a common complaint
OSCILLATION	X	X		X			X		X	Similar tests for complaints of "hum" and "noise." Align only if other tests indicate

Figure III–2
Universal Troubleshooting Chart

If the equipment is *unstable,* the trouble is OSCILLATION. The test methods given for this problem also apply where circuit operation is upset by an undesired signal such as *hum, noise,* or *interference.*

In most cases any of several test techniques may be used, depending upon the test instruments available and the type of equipment involved. Two general checks should be made, regardless of the trouble: *visual inspection* and *power supply verification* (incorrect voltages can cause a variety of problems). Where alignment is indicated by the chart in Figure III-2, it applies only to equipment with fixed tuned circuits.

Visual Inspection

Figure III-3 outlines in block diagram form the basic service procedure for all equipment, whether or not it contains ICs. The first step, of course, is a visual inspection of the equipment, watching for obvious physical or electrical damage. This, in itself, may permit a quick isolation of the defect. Then check the power supply. If the equipment uses batteries, check these. Substituting fresh ones, or changing to alkaline batteries, has often been known to solve the problem.

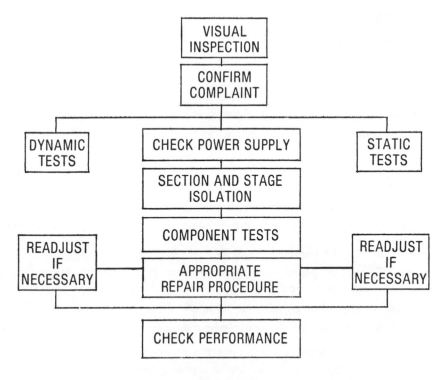

Figure III–3
Basic Troubleshooting Procedure

Confirm Complaint

You should always do this whenever you are called upon to troubleshoot somebody elsc's equipment. Your assessment of the problem may be quite different from the owner's, since yours is based upon knowledge and experience that he does not possess. Even when the equipment is your own, it is as well to shut it down, and then start all over again, to see whether the malfunction is repeated.

Resistance Checks

Point-to-point *resistance checks* are useful for isolating defects which cause a change in DC resistance values. However, remember that in equipment employing transistors misleading results can arise because of the direct resistive connection between a transistor's various electrodes. It will, therefore, be necessary to remove ICs or transistors before making resistance measurements. This is no problem if they are installed in sockets, but if they are soldered in place it is best to skip resistance tests until other tests that can be performed without removing these devices have been done. Use the following technique:

1. Turn off the power.

2. Discharge any large capacitors.

3. Unplug ICs or transistors. (If soldered, *only* from stage or stages to be checked.)

4. Using an ohmmeter, measure the DC resistance between each IC's or transistor's socket pin (or other connection point) and ground or V_{CC} bus.

5. Compared readings obtained with those given in the schematic diagram or service manual.

6. Ignore *minor* differences between "actual" and "expected" values. *Major* differences indicate trouble.

DC Voltage Analysis

DC voltage analysis is perhaps the oldest of all test techniques, and is valuable for isolating defects that cause a change in a circuit's operating voltages. The procedure is as follows:

1. Turn the power on and adjust any controls for normal operation.

2. Using a DC voltmeter, check voltages between each IC or transistor pin and ground, being careful to observe proper polarity.

3. Compare readings obtained with the schematic diagram or service manual.

4. Ignore *minor* discrepancies. *Major* differences indicate trouble.

In addition to DC voltage tests, *current measurements* are sometimes helpful for locating troubles. An overall current test may be made as a check on equipment efficiency. For this the meter is inserted in series with one of the power supply or battery leads.

Dynamic Tests

Dynamic tests are made under actual operating conditions. These tests may be used to isolate all kinds of troubles, from serious component defects causing a major change in the equipment operation to minor defects lowering circuit efficiency and performance, but permitting "nearly normal" operation. Dynamic tests are also valuable for tracking down intermittent defects.

Signal tracing is the most powerful dynamic technique. It consists of tracing a signal stage by stage as it passes through the equipment. The instrument used to follow the signal may be an AC voltmeter, a signal tracer, or an oscilloscope. Generally, a signal tracer (or a receiver's own loudspeaker) is used to follow signals through receivers and audio amplifiers, while an oscilloscope is superior for most other items. An oscilloscope for use in RF circuits must have the necessary bandwidth, or have a demodulator probe.

To use the signal-tracing technique, proceed as follows, referring to Figure III-4:

1. Turn the power on.

2. Apply an adequate test signal to the equipment's input. For a receiver, this signal may be obtained from a signal injector, an RF signal generator, or by tuning in a local station. For an audio amplifier, a suitable signal may be obtained from a record player, tuner, or audio

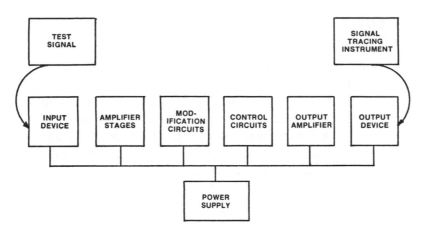

Figure III–4
Signal Tracing Technique

signal generator. In other cases, depending upon the nature of the equipment being tested, the signal may be obtained from a pulse generator, square wave generator, sawtooth oscillator, or other device. If the equipment generates its own signal (for instance, a radio transmitter) a separate signal source may not be needed.

3. Make sure that connecting the signal source does not disturb the circuit. If necessary, insert a DC blocking capacitor, matching pad, or other device as necessary, in series with the input.

4. Make sure the test signal does not overload the equipment. Always use the least signal amplitude needed for usable output.

5. Use a signal tracing instrument (signal tracer, oscilloscope, etc.) to check the relative amplitude and quality of the signal at the input and output of each stage, starting with the last and working forward, as shown in Figure III-4. Depending on the instrument used, the signal may be: (a) heard in a loudspeaker; (b) observed as the closure of a ''tuning eye''; (c) indicated by the deflection of a meter pointer; or (d) seen as a waveform on a CRT.

6. The test signal should be modified by each stage. For instance, an amplifier should increase its amplitude; a clipper should remove a portion of it.

7. If the signal is changed in an unexpected fashion (for example, the amplitude drops where it should have increased), trouble is indicated, and the defective stage has been isolated.

8. Where necessary, change the type of pickup (probe) as the signal is followed through the various stages. For example, an RF detector probe is used for checking the RF and IF stages of a receiver, a direct probe for audio stages.

Waveform analysis is used to investigate changes in equipment performance that are minor rather than major or catastrophic—a deterioration in frequency response, for instance. Here, the oscilloscope is the instrument to use. The test signal may be a sine wave, square wave, pulse, AM or FM carrier, or other signal appropriate to the equipment. A dual-trace scope is ideal for this, because you can show both the input and the output signals on its screen at the same time, making comparison very easy.

Signal injection is a technique that works very well with RF, IF, and AF circuits. The *signal injector* is a pocket-sized battery-operated device that applies a signal to the input of any stage in a radio receiver. The signal passes through the stages between the point of injection and the final stage, resulting in an audible tone in the loudspeaker. The usual procedure is to start at the final stage, and work forward stage by stage until you reach the antenna. Each time you move another stage forward, the signal receives that much more amplification, which is normal. However, if it becomes weaker, or disappears, you have located the stage where the trouble exists.

If you do not have a signal injector you can use a signal generator, or even your finger, because your body contact applied to the grid of a tube or the base of a transistor will disturb the circuit enough to cause an audible pop or click in the speaker.

Brute force means lightly tapping, wiggling, or otherwise physically manipulating the circuitry to reveal the source of an intermittent symptom such as a defective connection. Some types of intermittent behavior are temperature related. You may have to get results by warming or cooling. The safest way of warming the equipment is to place it in a box with a 40-watt light bulb. Cooling is done by spraying the suspected part or parts with freon from a spray can.

Part substitution is the technique of replacing a suspected part with another that you know to be good, to see if this clears up the problem. It is a useful practice with vacuum tubes, and components such as electrolytic capacitors, which do not always give a bad indication on a test instrument.

Appendix

COIL DATA

Single-Layer Wound Coil—The formulas for winding a radio-frequency, air-core coil with one layer of closely wound enamel wire are:

$$L = \frac{(rN)^2}{9r + 10l} \tag{1}$$

$$N = \frac{\sqrt{L(9r + 10l)}}{r} \tag{2}$$

where L = self-inductance of coil, in microhenries,
N = total number of turns,
r = mean radius of coil, in inches, or centimeters,
l = length of coil in inches, or centimeters.

Multi-Layer Wound Coil—The formulas for winding a radio-frequency, air-core coil with more than one layer of closely wound enamel wire are (3) below and (2) above:

$$L = \frac{0.8\,(rN)^2}{6r + 9l + 10b} \tag{3}$$

where L = self-inductance of coil, in microhenries,
N = total number of turns,
l = length of coil in inches, or centimeters,
b = (outside radius) – (inside radius), in inches, or centimeters,
r = mean radius of coil, in inches, or centimeters.

Data for these calculations are given in the following table for copper wire sizes from AWG#10 to AWG#40 (Figure A–1).

Inductance, Capacitance, and Reactance Charts—The charts in Figures A-2, A-3, and A-4 may be used to determine unknown values of frequency, inductance, capacitance, and reactance. Any one unknown value may be found if two of the others are known. Using the correct chart for the frequency, lay a straightedge across it so it passes through the two known values, and find the required value where it intersects the appropriate scale.

ENAMELED COPPER WIRE TABLE

Gage (AWG/BS)	Nominal Diameter (in)	Nominal Diameter (mm)	Turns/inch	Turns/cm	Ohms/1000'	Ohms/100 m
10	0.1019	2.588	9.6	3.8	0.9989	0.3277
11	0.09074	2.305	10.7	4.2	1.260	0.4134
12	0.08081	2.053	12.0	4.7	1.588	0.5210
13	0.07196	1.828	13.5	5.3	2.003	0.6572
14	0.06408	1.628	15.0	5.9	2.525	0.8284
15	0.05707	1.450	16.8	6.6	3.184	1.045
16	0.05082	1.291	18.9	7.4	4.016	1.318
17	0.04526	1.150	21.2	8.3	5.064	1.661
18	0.04030	1.024	23.6	9.3	6.385	2.095
19	0.03589	0.9116	26.4	10.4	8.051	2.641
20	0.03196	0.8118	29.4	11.8	10.15	3.330
21	0.02846	0.7229	33.1	13.03	12.80	4.200
22	0.02535	0.6439	37.0	14.6	16.14	5.300
23	0.02257	0.5733	41.3	16.3	20.36	6.680
24	0.02010	0.5105	46.3	18.2	25.67	8.422
25	0.01790	0.4547	51.7	20.4	32.37	10.62
26	0.01594	0.4049	58.0	22.8	40.81	13.39
27	0.01420	0.3607	64.9	25.6	51.47	16.89
28	0.01264	0.3211	72.7	28.7	64.90	21.29
29	0.01126	0.2860	81.6	32.1	81.83	26.88
30	0.01003	0.2548	90.5	35.6	103.2	33.86
31	0.008928	0.2268	101	39.8	130.1	42.68
32	0.007950	0.2019	113	44.5	164.1	53.84
33	0.007080	0.1798	127	50.0	206.9	67.88
34	0.006305	0.1601	143	56.3	260.9	85.60
35	0.005615	0.1426	158	62.2	329.0	107.9
36	0.005000	0.1270	175	68.9	414.8	136.1
37	0.004453	0.1131	198	78.0	523.1	171.6
38	0.003965	0.1007	224	88.2	659.6	216.4
39	0.003531	0.08969	248	97.6	831.8	272.9
40	0.003145	0.07988	282	111.0	1049.0	344.2

Figure A-1

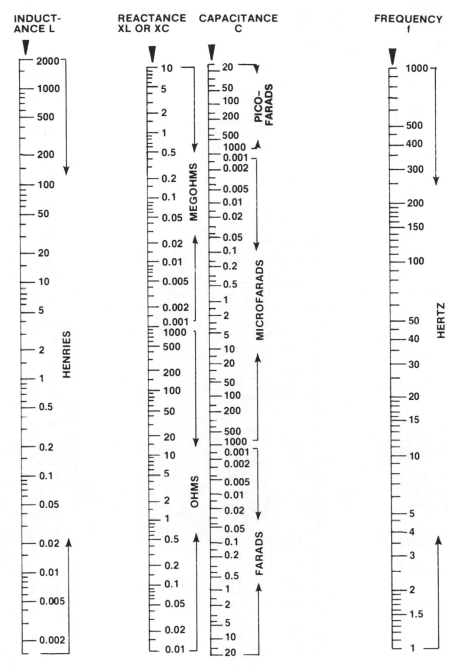

Figure A-2

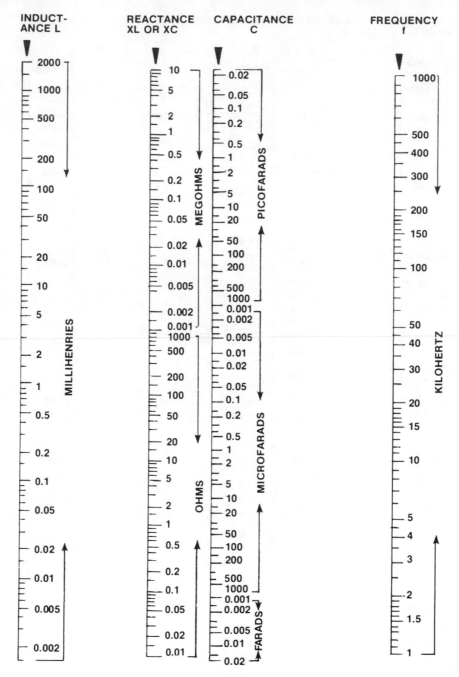

Figure A-3

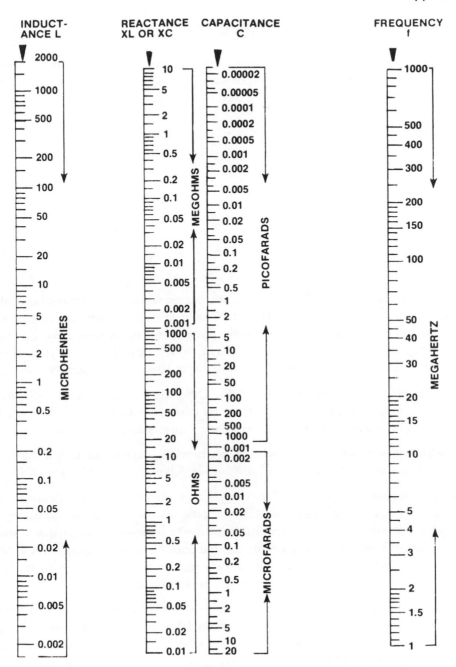

Figure A-4

COLOR CODING

Fixed non-metallic resistors and some other components are color-coded to indicate their values instead of having the information printed or stamped on them. The colors used and their numerical significance are as follows:

Black	0
Brown	1
Red	2
Orange	3
Yellow	4
Green	5
Blue	6
Violet	7
Gray	8
White	9

On a *composition resistor* there are three or four colored bands around the resistor body, adjacent to one end. Starting with the one nearest the end, the first two colors are the significant figures of the resistor value in ohms, and the third is the power of ten of the multiplier. For example, blue (6), red (2), and yellow (4) should be read as 62×10^4, or 620 kilohms. If the third band is gray (or silver), or white (or gold), the multiplier is 10^{-2}, or 10^{-1}, respectively.

If there is no fourth band the tolerance is 20%. Otherwise, silver denotes 10%, gold 5%. Other tolerances may be indicated by using the applicable colors, but these are seldom encountered as most precision resistors have their values stamped or printed on them.

Some *film-type* resistors also may be color-coded, in which case an additional color may be added in front, to provide for three significant figures, followed by a multiplier and tolerance.

Capacitors, for the most part, are not color-coded. Exceptions are some *ceramic* and *mica* types. Ceramic *tubular* capacitors may be coded in the same way as composition resistors, except that an additional color is added in front of the others to indicate the *temperature characteristic.* The colored bands may also be replaced by colored dots. Ceramic *disks* have these arranged around the edge, to be read clockwise. Some very small tubular capacitors have only three dots, which give two significant figures and the multiplier only.

Mica capacitors are now mostly not color-coded. However, the flat rectangular molded silvered-mica type may employ an arrangement of six colored dots arranged three on each side of an arrow, or some mark to indicate which way to read them. With the arrow pointing to the right the upper three dots (reading from left to right) indicate: EIA standard (this is a white dot, which may be omitted) and two significant figures. The bottom row (reading from right to left) gives the multiplier, tolerance, and type.

The significant figures always give the value in picofarads. The temperature characteristic (ceramic capacitors), or type (mica capacitors), is denoted as follows:

COLOR	TEMPERATURE CHARACTERISTIC	TYPE (MFR'S SPEC.)
Black	NPO	A
Brown	N033	B
Red	N075	C
Orange	N150	D
Yellow	N220	E
Green	N330	
Blue	N470	
Violet	N750	

Diodes used in signal circuits are often color-coded because of their small size. Diode designations are always given, for example, in the form 1N914A. The 1N- is common to all and does not have to be indicated; therefore, the color bands denote the figures and letter following. Suffix letters are coded as follows:

Black	(no suffix)
Brown	A
Red	B
Orange	C
Yellow	D
Green	E
Blue	F
Violet	G
Gray	H
White	J

The colored bands are grouped at the cathode end of the diode, and should be read from that end.

Transformer leads are color-coded as follows:

POWER TRANSFORMERS

Primary (tapped)   *Secondary*

Black (common)

Black/yellow (centertap)   High voltage { Red
                                                  Red/yellow
                                                  (centertap)
Black/red   Red

POWER TRANSFORMERS (cont'd.)

Primary (untapped)	Rectifier filament	{ Yellow Yellow/blue (centertap) Yellow
Two blacks leads		
	Amplifier filament #1	{ Green Green/yellow (centertap) Green
	Amplifier filament #2	{ Brown Brown/yellow (centertap) Brown
	Amplifier filament #3	{ Slate Slate/yellow (centertap) Slate

IF TRANSFORMERS

Primary	*Secondary*
Blue (plate)	Green (grid or diode)
	Violet (full-wave diode)
Red (B +)	White (grid or diode return, AVC, or ground)

AUDIO & OUTPUT TRANSFORMERS

Primary	*Secondary*
Blue (plate)	Green (grid or voice coil)
Red (B +)	Black (return or voice coil)
*Blue or brown (plate)	*Green or yellow (grid)

CONVERSION FACTORS

See Also SI Units (Metric System).

To convert	to	multiply by		conversely, multiply by	
abampere	ampere	1.00	$\times 10^{1}$	1.00	$\times 10^{-1}$
abcoulomb	coulomb	1.00	$\times 10^{1}$	1.00	$\times 10^{-1}$
abfarad	farad	1.00	$\times 10^{9}$	1.00	$\times 10^{-9}$
abhenry	henry	1.00	$\times 10^{-9}$	1.00	$\times 10^{9}$
abmho	siemens	1.00	$\times 10^{9}$	1.00	$\times 10^{-9}$
abohm	ohm	1.00	$\times 10^{-9}$	1.00	$\times 10^{9}$
abvolt	volt	1.00	$\times 10^{-8}$	1.00	$\times 10^{8}$
ampere (1948)	ampere	9.998 35	$\times 10^{-1}$	1.000 17	$\times 10^{0}$
angstrom	meter	1.00	$\times 10^{-10}$	1.00	$\times 10^{10}$
atmosphere	pascal	1.013 25	$\times 10^{5}$	9.869 23	$\times 10^{-6}$
bar	pascal	1.00	$\times 10^{5}$	1.00	$\times 10^{-5}$
barn	meter^{2}	1.00	$\times 10^{-28}$	1.00	$\times 10^{28}$
British Thermal Unit (mean)	joule	1.055 87	$\times 10^{3}$	9.470 86	$\times 10^{-4}$
calorie (thermochemical)	joule	4.184	$\times 10^{0}$	2.390	$\times 10^{-1}$
calorie (kilogram, thermochemical)	joule	4.184	$\times 10^{3}$	2.390	$\times 10^{-4}$
circular mil	meter^{2}	5.067 074 8	$\times 10^{-10}$	1.973 525 2	$\times 10^{9}$
coulomb (1948)	coulomb	9.998 35	$\times 10^{-1}$	1.000 17	$\times 10^{0}$
cubic foot	meter^{3}	2.831 69	$\times 10^{-2}$	3.531 46	$\times 10^{1}$
cubic inch	meter^{3}	1.638 71	$\times 10^{-5}$	6.102 36	$\times 10^{4}$
cubic yard	meter^{3}	7.645 55	$\times 10^{-1}$	1.307 95	$\times 10^{0}$
curie	becquerel	3.70	$\times 10^{10}$	2.70	$\times 10^{-11}$
degree	radian	1.745 329 2	$\times 10^{-2}$	5.729 577 9	$\times 10^{1}$
dyne	newton	1.00	$\times 10^{-5}$	1.00	$\times 10^{5}$
farad (1948)	farad	9.995 05	$\times 10^{-1}$	1.000 50	$\times 10^{0}$
faraday (carbon 12)	coulomb	9.648 70	$\times 10^{4}$	1.036 41	$\times 10^{-5}$
fathom	meter	1.828 8	$\times 10^{0}$	5.468 07	$\times 10^{-1}$
fermi	meter	1.00	$\times 10^{-15}$	1.00	$\times 10^{15}$
fluid ounce (U.S.)	meter^{3}	2.957 353	$\times 10^{-5}$	3.381 402	$\times 10^{4}$
foot	meter	3.048	$\times 10^{-1}$	3.281	$\times 10^{0}$
foot-candle	lumen/meter^{2}	1.076 391 0	$\times 10^{1}$	9.290 304 3	$\times 10^{-2}$
foot-lambert	candela/meter^{2}	3.426 259	$\times 10^{0}$	2.918 635	$\times 10^{-1}$
gal	meter/second^{2}	1.00	$\times 10^{-2}$	1.00	$\times 10^{2}$

CONVERSION FACTORS (continued)

To convert	to	multiply by		conversely, multiply by	
gallon (U.S. liquid)	meter^3	3.785 411 8	$\times 10^{-3}$	2.641 172 1	$\times 10^2$
gamma	tesla	1.00	$\times 10^{-9}$	1.00	$\times 10^9$
gauss	tesla	1.00	$\times 10^{-4}$	1.00	$\times 10^4$
gilbert	ampere turn	7.957 747 2	$\times 10^{-1}$	1.256 637 0	$\times 10^0$
grain	kilogram	6.479 891	$\times 10^{-5}$	1.543 236	$\times 10^4$
henry (1948)	henry	1.000 495	$\times 10^0$	9.995 052	$\times 10^{-1}$
horsepower (electric)	watt	7.46	$\times 10^2$	1.34	$\times 10^{-3}$
inch	meter	2.54	$\times 10^{-2}$	3.94	$\times 10^1$
inch of mercury (32° F)	pascal	3.386 389	$\times 10^3$	2.952 998	$\times 10^{-4}$
joule (1948)	joule	1.000 165	$\times 10^0$	9.998 350	$\times 10^{-1}$
knot	meter/second	5.144 444 4	$\times 10^{-1}$	1.943 844 5	$\times 10^0$
lambert	candela/meter^2	3.183 098 8	$\times 10^3$	3.141 592 7	$\times 10^{-4}$
liter	meter^3	1.00	$\times 10^{-3}$	1.00	$\times 10^3$
lux	lumen/meter^2	1.00	$\times 10^0$	1.00	$\times 10^0$
maxwell	weber	1.00	$\times 10^{-8}$	1.00	$\times 10^8$
mho	siemens	1.00	$\times 10^0$	1.00	$\times 10^0$
micron	meter	1.00	$\times 10^{-6}$	1.00	$\times 10^6$
mil	meter	2.540	$\times 10^{-5}$	3.937	$\times 10^4$
mile (U.S. statute)	meter	1.609 344	$\times 10^3$	6.213 712	$\times 10^{-4}$
mile (nautical)	meter	1.852	$\times 10^3$	5.400	$\times 10^{-4}$
millibar	pascal	1.00	$\times 10^2$	1.00	$\times 10^{-2}$
millimeter of mercury (0° C)	pascal	1.333 224	$\times 10^2$	7.500 615	$\times 10^{-3}$
neper	decibel	8.686	$\times 10^0$	1.151	$\times 10^{-1}$
newton/meter^2	pascal	1.00	$\times 10^0$	1.00	$\times 10^0$
oersted	ampere/meter	7.957 747 2	$\times 10^1$	1.256 637 0	$\times 10^{-2}$
ohm (1948)	ohm	1.000 495	$\times 10^0$	9.995 052	$\times 10^{-1}$
ounce force (avoirdupois)	newton	2.780 138 5	$\times 10^{-1}$	3.596 943 1	$\times 10^0$
ounce mass (avoirdupois)	kilogram	2.834 952 3	$\times 10^{-2}$	3.215 074 6	$\times 10^1$
pieze	pascal	1.00	$\times 10^3$	1.00	$\times 10^{-3}$
pint (U.S. liquid)	meter^3	4.731 764 7	$\times 10^{-4}$	2.113 376 4	$\times 10^3$
poise	newton second /meter^2	1.00	$\times 10^{-1}$	1.00	$\times 10^1$

CONVERSION FACTORS (continued)

To convert	to	multiply by		conversely, multiply by	
pound force (avoirdupois)	newton	4.448 221 6	$\times 10^{0}$	2.248 089 4	$\times 10^{-1}$
pound mass (avoirdupois)	kilogram	4.535 923 7	$\times 10^{-1}$	2.204 622 6	$\times 10^{0}$
poundal	newton	1.382 549 5	$\times 10^{-1}$	7.233 014 0	$\times 10^{0}$
quart (U.S. liquid)	liter	9.463 529 5	$\times 10^{-1}$	1.056 688 2	$\times 10^{0}$
quart (U.S. liquid)	meter^{3}	9.463 529 5	$\times 10^{-4}$	1.056 688 2	$\times 10^{3}$
rad	joule/kilogram	1.00	$\times 10^{-2}$	1.00	$\times 10^{2}$
rayleigh	1/second meter^{2}	1.00	$\times 10^{10}$	1.00	$\times 10^{-10}$
roentgen	coulomb/ kilogram	2.579 76	$\times 10^{-4}$	3.876 33	$\times 10^{3}$
rutherford	becquerel	1.00	$\times 10^{6}$	1.00	$\times 10^{-6}$
slug	kilogram	1.459 390 3	$\times 10^{1}$	6.852 176 5	$\times 10^{-2}$
square inch	meter^{2}	6.452	$\times 10^{-4}$	1.550	$\times 10^{3}$
square foot	meter^{2}	9.290	$\times 10^{-2}$	1.076	$\times 10^{1}$
square yard	meter^{2}	8.361	$\times 10^{-1}$	1.196	$\times 10^{0}$
statampere	ampere	3.335 640	$\times 10^{-10}$	2.997 925	$\times 10^{9}$
statcoulomb	coulomb	3.335 640	$\times 10^{-10}$	2.997 925	$\times 10^{9}$
statfarad	farad	1.112 650	$\times 10^{-12}$	8.987 554	$\times 10^{11}$
stathenry	henry	8.987 554	$\times 10^{11}$	1.112 650	$\times 10^{-12}$
statmho	siemens	1.112 650	$\times 10^{-12}$	8.987 554	$\times 10^{11}$
statohm	ohm	8.987 554	$\times 10^{11}$	1.112 650	$\times 10^{-12}$
statvolt	volt	2.997 925	$\times 10^{2}$	3.335 640	$\times 10^{-3}$
stilb	candela/meter^{2}	1.00	$\times 10^{4}$	1.00	$\times 10^{-4}$
torr (0°C)	pascal	1.333 22	$\times 10^{2}$	7.500 64	$\times 10^{-3}$
unit pole	weber	1.256 637	$\times 10^{-7}$	7.957 748	$\times 10^{6}$
volt (1948)	volt	1.000 330	$\times 10^{0}$	9.996 701	$\times 10^{-1}$
watt (1948)	watt	1.00 165	$\times 10^{0}$	9.998 350	$\times 10^{-1}$
yard	meter	9.144	$\times 10^{-1}$	1.094	$\times 10^{0}$

Temperature conversion formulas

Fahrenheit to Celsius:

$$t_c = \left(\frac{5}{9}\right)(t_f - 32)$$

Celsius to Fahrenheit:

$$t_f = \frac{9t_c}{5} + 32$$

Celsius to Kelvin: $\qquad$ $t_k = t_c + 273.15$

Kelvin to Celsius: $\qquad$ $t_c = t_k - 273.15$

ELECTRICAL EQUATIONS

The following equations are those required most often in the field of electronics. The meanings of the symbols used are listed below (numerical subscripts are added when the same symbol is used for more than one quantity in the same formula).

A $\quad$ = length of side adjacent to θ in the right triangle, in same units as other sides

B $\quad$ = susceptance, in siemens

C $\quad$ = capacitance, in farads

D $\quad$ = dissipation factor

d $\quad$ = thickness of dielectric (spacing of plates), in centimeters

dB $\quad$ = decibels

E $\quad$ = potential, in volts

F $\quad$ = temperature, in degrees Fahrenheit

f $\quad$ = frequency, in hertz

G $\quad$ = conductance, in siemens

H $\quad$ = length of hypotenuse (side opposite right angle) in right triangle, in same units as other sides

I $\quad$ = current, in amperes

K $\quad$ = dielectric constant; coupling coefficient: or temperature, in kelvins

L $\quad$ = self inductance, in henries

M $\quad$ = mutual inductance, in henries

N $\quad$ = number of plates; or number of turns

O $\quad$ = length of side opposite to θ in right triangle, in same units as other sides

P $\quad$ = power, in watts

p.f. $\quad$ = power factor

Q $\quad$ = figure of merit; or quantity of electricity stored, in coulombs

R $\quad$ = resistance, in ohms

S $\quad$ = area of one plate of capacitor, in square centimeters

X $\quad$ = reactance, in ohms

X_c $\quad$ = capacitive reactance, in ohms

ELECTRICAL EQUATIONS (cont'd.)

X_l = inductive reactance, in ohms

Y = admittance, in siemens

Z = impedance

δ = 90-θ degrees

θ = phase angle, in degrees; angle, in degrees, in right triangle, whose sine, cosine, tangent, etc., is required.

λ = wavelength, in meters

π = 3.1416...

Admittance: (1) $Y = \dfrac{1}{\sqrt{R^2 + x^2}}$

(2) $Y = \dfrac{1}{Z}$

(3) $Y = \sqrt{G^2 + B^2}$

Average value: (1) Average value = 0.637 × peak value

(2) Average value = 0.900 × r.m.s. value

Capacitance: (1) Capacitors in parallel:

$C = C_1 + C_2 + C_3 \ldots$ etc.

(2) Capacitors in series:

$C = \dfrac{1}{\dfrac{1}{C_1} + \dfrac{1}{C_2} + \dfrac{1}{C_3} \ldots \text{ etc.}}$

(3) Two capacitors in series:

$C = \dfrac{C_1 C_2}{C_1 + C_2}$

(4) Capacitance of capacitor:

$C = 0.0885 \ \dfrac{KS(N-1)}{d}$

(5) Quantity of electricity stored:

$Q = CE$

Conductance: (1) $G = \dfrac{1}{R}$

ELECTRICAL EQUATIONS (cont'd.)

(2) $G = \dfrac{I}{E}$

(3) $G_{total} = G_1 + G_2 + G_3 \ldots$
(resistors in parallel)

(4) $I_{total} = EG_{total}$

(5) $I_2 = \dfrac{I_{total}G_2}{G_1 + G_2 + G_3 \ldots \text{etc.}}$ (current in R_2)

Cosecant:

(1) $\csc \theta = \dfrac{H}{O}$

(2) $\csc \theta = \sec (90 - \theta)$

(3) $\csc \theta = \dfrac{1}{\sin \theta}$

Cosine:

(1) $\cos \theta = \dfrac{A}{H}$

(2) $\cos \theta = \sin (90 - \theta)$

(3) $\cos \theta = \dfrac{1}{\sec \theta}$

Cotangent:

(1) $\cot \theta = \dfrac{A}{O}$

(2) $\cot \theta = \tan (90 - \theta)$

(3) $\cot \theta = \dfrac{1}{\tan \theta}$

Decibel:

(1) $dB = 10 \log \dfrac{P_1}{P_2}$

(2) $dB = 20 \log \dfrac{E_1}{E_2}$ (source and load impedance equal)

(3) $dB = 20 \log \dfrac{I_1}{I_2}$ (source and load impedance equal)

(4) $dB = 20 \log \dfrac{E_1 \sqrt{Z_2}}{E_2 \sqrt{Z_1}}$ (source and load impedances unequal)

(5) $dB = 20 \log \dfrac{I_1 \sqrt{Z_1}}{I_2 \sqrt{Z_2}}$ (source and load impedances unequal)

Frequency:

(1) $f = \dfrac{3 \times 10^8}{\lambda}$

(2) $f = \dfrac{1}{2\pi\sqrt{LC}}$

ELECTRICAL EQUATIONS (cont'd.)

Impedance:

(1) $Z = \sqrt{R^2 + X^2}$

(2) $Z = \sqrt{G^2 + B^2}$

(3) $Z = \dfrac{R}{\cos \theta}$

(4) $Z = \dfrac{X}{\sin \theta}$

(5) $Z = \dfrac{E}{I}$

(6) $Z = \dfrac{P}{I^2 \cos \theta}$

(7) $Z = \dfrac{E^2 \cos \theta}{P}$

Inductance:

(1) Inductors in series: $L = L_1 + L_2 + L_3 \ldots$ etc.

(2) Inductors in parallel:

$$L = \dfrac{1}{\dfrac{1}{L_1} + \dfrac{1}{L_2} + \dfrac{1}{L_3} \ldots \text{etc.}}$$

(3) Two inductors in parallel: $L = \dfrac{L_1 L_2}{L_1 + L_2}$

(4) Coupled inductances in series with fields aiding: $L = L_1 + L_2 + 2M$

(5) Coupled inductances in series with fields opposing: $L = L_1 + L_2 - 2M$

(6) Coupled inductances in parallel with fields aiding:

$$L = \dfrac{1}{\dfrac{1}{L_1 + M} + \dfrac{1}{L_2 + M}}$$

(7) Coupled inductances in parallel with fields opposing:

$$L = \dfrac{1}{\dfrac{1}{L_1 - M} + \dfrac{1}{L_2 - M}}$$

(8) Mutual induction of two RF coils with fields interacting:

$$M = \dfrac{L_1 - L_2}{4}$$

ELECTRICAL EQUATIONS (cont'd.)

where L_1 - total inductance of both coils with fields aiding
where L_2 - total inductance of both coils with fields opposing

(9)   Coupling coefficient of two RF coils inductively coupled so as to give transformer action:

$$K = \frac{M}{\sqrt{L_1 L_2}}$$

(Other coil formulas are given under *Coil Data*.)

Meter formulas:

(1)   Ohms per volt $= \dfrac{1}{I}$

(I = full-scale current in amperes)

(2)   Meter resistance: $R_{meter} = R_{rheostat}$

(The meter is connected in series with a battery and a rheostat, and the rheostat is adjusted until the meter reads full scale. A second rheostat is then connected in parallel with the meter and adjusted until the meter reads half scale. The resistance of the second rheostat will equal that of the meter.)

(3)   Current shunt: $R = \dfrac{R_{meter}}{N - 1}$

where N is the new full-scale reading divided by the original full-scale reading (both in the same units).

(4)   Voltage multiplier:

$$R = \frac{\text{Full-scale reading required}}{\text{Full-scale current of meter}} - R_{meter}$$

where reading is in volts and current in amperes.

Ohm's law formulas for DC circuits:

(1) $I = \dfrac{E}{R}$

(2) $I = \sqrt{\dfrac{P}{R}}$

(3) $I = \dfrac{P}{E}$

(4) $R = \dfrac{E}{I}$

(5) $R = \dfrac{P}{I^2}$

ELECTRICAL EQUATIONS (cont'd.)

(6) $\quad R = \dfrac{E^2}{P}$

(7) $\quad E = IR$

(8) $\quad E = \dfrac{P}{I}$

(9) $\quad E = \sqrt{PR}$

(10) $\quad P = I^2R$

(11) $\quad P = EI$

(12) $\quad P = \dfrac{E^2}{R}$

Ohm's law formulas for AC circuits:

(1) $\quad I = \dfrac{E}{Z}$

(2) $\quad I = \sqrt{\dfrac{P}{Z \cos \theta}}$

(3) $\quad I = \dfrac{P}{E \cos \theta}$

(4) $\quad Z = \dfrac{E}{I}$

(5) $\quad Z = \dfrac{P}{I^2 \cos \theta}$

(6) $\quad Z = \dfrac{E^2 \cos \theta}{P}$

(7) $\quad E = IZ$

(8) $\quad E = \dfrac{P}{I \cos \theta}$

(9) $\quad E = \sqrt{\dfrac{PZ}{\cos \theta}}$

(10) $\quad P = I^2Z \cos \theta$

(11) $\quad P = IE \cos \theta$

(12) $\quad P = \dfrac{E^2 \cos \theta}{Z}$

Peak value:
(1) Peak value = 1.414 × r.m.s. value
(2) Peak value = 1.570 × average value

Peak-to-peak value:
(1) P-P value = 2.828 × r.m.s. value
(2) P-P value = 3.140 × average value

ELECTRICAL EQUATIONS (cont'd.)

Phase angle: $\theta = \text{arc tan} \dfrac{X}{R}$

Power factor:
(1) p.f. $= \cos\theta$
(2) $D = \cot\theta$

Q (figure of merit):
(1) $Q = \tan\theta$

(2) $Q = \dfrac{X}{R}$

Reactance:
(1) $X_L = 2\pi fL$

(2) $X_C = \dfrac{1}{2\pi fC}$

Resistance:
(1) Resistors in series: $R = R_1 + R_2 + R_3 \ldots$ etc.

(2) Resistors in parallel:

$$R = \dfrac{1}{\dfrac{1}{R_1} + \dfrac{1}{R_2} + \dfrac{1}{R_3} \ldots \text{etc.}}$$

(3) Two resistors in parallel: $R = \dfrac{R_1 R_2}{R_1 + R_2}$

Resonance:
(1) $f = \dfrac{1}{2\pi\sqrt{LC}}$

(2) $L = \dfrac{1}{4\pi^2 f^2 C}$

(3) $C = \dfrac{1}{4\pi^2 f^2 L}$

Right triangle:
(1) $\sin\theta = \dfrac{O}{H}$

(2) $\cos\theta = \dfrac{A}{H}$

(3) $\tan\theta = \dfrac{O}{A}$

(4) $\csc\theta = \dfrac{H}{O}$

(5) $\sec\theta = \dfrac{H}{A}$

ELECTRICAL EQUATIONS (cont'd.)

(6) $\cot \theta = \dfrac{A}{O}$

Root-mean-square value:

(1)   R.m.s. value $= 0.707 \times$ peak value

(2)   R.m.s. value $= 1.111 \times$ average value

Secant:

(1) $\sec \theta = \dfrac{H}{A}$

(2) $\sec \theta = \csc (90 - \theta)$

(3) $\sec \theta = \dfrac{1}{\cos \theta}$

Sine:

(1) $\sin \theta = \dfrac{O}{H}$

(2) $\sin \theta = \cos (90 - \theta)$

(3) $\sin \theta = \dfrac{1}{\csc \theta}$

Susceptance:

(1) $B = \dfrac{X}{R^2 + X^2}$

(2) $B = \dfrac{1}{X}$

(3) $B = B_1 + B_2 + B_3 \ldots$ etc.

Tangent:

(1) $\tan \theta = \dfrac{O}{A}$

(2) $\tan \theta = \cot (90 - \theta)$

(3) $\tan \theta = \dfrac{1}{\cot \theta}$

Temperature:

(1) $C = 0.556F - 17.8$

(2) $F = 1.8C + 32$

(3) $K = C + 273$

Transformer ratio:

$\dfrac{N_p}{N_s} = \dfrac{E_p}{E_s} = \dfrac{I_s}{I_p} = \sqrt{\dfrac{Z_p}{Z_s}}$

(subscript p $=$ primary; subscript s $=$ secondary)

Wavelength:

$\lambda = \dfrac{3 \times 10^8}{f}$

ELECTROMAGNETIC SPECTRUM

Frequency (hertz)		Wavelength (meters)		Radioactive and Radio Waves	Ultraviolet, Visible, Infrared and Audio Waves
100 EHz	$(\times 10^{18})$	3 pm	$(\times 10^{-12})$	Gamma rays	
10 EHz	"	30 pm	"		
1 EHz	"	300 pm	"	(hard)	
100 PHz	$(\times 10^{15})$	3 nm	$(\times 10^{-9})$	X rays	
10 PHz	"	30 nm	"	(soft)	Ultraviolet rays
1 PHz	"	300 nm	"		
100 THz	$(\times 10^{12})$	3 μm	$(\times 10^{-6})$		Visible light rays
10 THz	"	30 μm	"		Infrared rays
1 THz	"	300 μm	"	12	
100 GHz	$(\times 10^{9})$	3 mm	$(\times 10^{-3})$	EHF - 11	
10 GHz	"	30 mm	"	SHF - 10	
1 GHz	"	300 mm	"	UHF - 9	
100 MHz	$(\times 10^{6})$	3 m	$(\times 10^{0})$	VHF - 8	
10 MHz	"	30 m	"	HF - 7	
1 MHz	"	300 m	"	MF - 6	
100 kHz	$(\times 10^{3})$	3 km	$(\times 10^{3})$	LF - 5	
10 kHz	"	30 km	"	VLF - 4	
1 kHz	"	300 km	"	VF - 3	Audio waves
100 Hz	$(\times 10^{0})$	3 Mm	$(\times 10^{6})$	ELF - 2	
10 Hz	"	30 Mm	"		
1 Hz	"	300 Mm	"		

GRAPHIC SYMBOLS

AMPLIFIER (A)

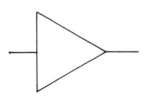

1. General:
   Symbol represents any method
   of amplification (vacuum tube,
   semiconductor, etc.).

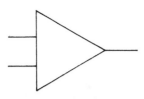

2. Amplifier with two inputs

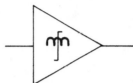

3. Magnetic amplifier

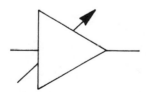

4. Amplifier with adjustable gain

ANTENNA (E)

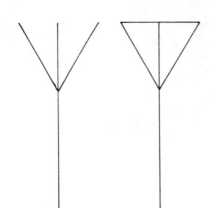

1. General

GRAPHIC SYMBOLS (cont'd.)

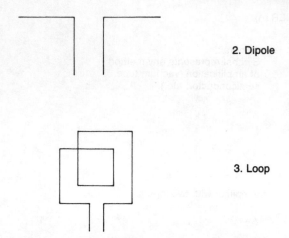

2. Dipole

3. Loop

AUDIBLE SIGNALING DEVICE (LS)

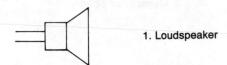

1. Loudspeaker

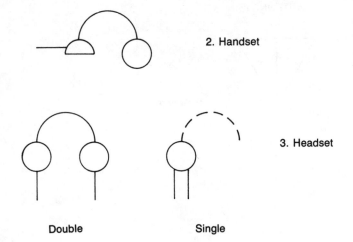

2. Handset

3. Headset

Double   Single

GRAPHIC SYMBOLS (cont'd.)

BATTERY (BT)

1. One cell; also used for generalized direct current source. The long line is always positive, but polarity may be indicated in addition.

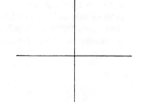

2. Multicell

CABLE, CONDUCTOR, WIRING (W)

1. General, any guided path

2. Crossing of paths or conductors, not connected

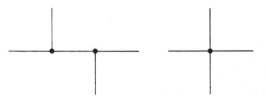

3. Junction of paths; dots are optional unless right-hand junction is used.

Preferred   Only if required by layout considerations

GRAPHIC SYMBOLS (cont'd.)

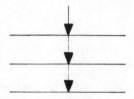

4. Associated conductors: general (shown with three conductors)

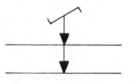

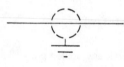

5. Twisted pair: twist symbol may be replaced with letter P for pair, or T for triple.

6. Shielded conductor, with shield grounded

7. Coaxial cable; the broken line is not part of the symbol but indicates where outer conductor connection to another symbol is made.

8. Circular waveguide

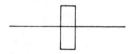

9. Rectangular waveguide

GRAPHIC SYMBOLS (cont'd.)

CAPACITOR (C)

Style 1   Style 2

1. Capacitor; the curved or modified electrode indicates the outside, low potential or movable element.

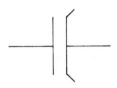

Style 1 modified
to identify electrode

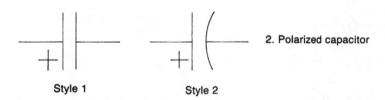

Style 1   Style 2

2. Polarized capacitor

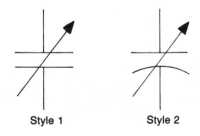

Style 1   Style 2

3. Adjustable or variable capacitor. If mechanical linkage of more than one unit is to be shown, the tails of the arrows are joined by a dashed line.

GRAPHIC SYMBOLS (cont'd.)

CIRCUIT BREAKER (CB)

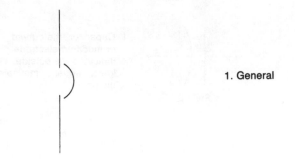

1. General

CONNECTOR: FEMALE (J), MALE (P)

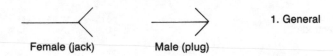

1. General

Female (jack)   Male (plug)

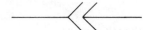

2. Male and female connectors engaged

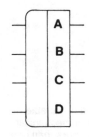

3. Alternative way of showing multiple connector. Plug is on the left, jack on the right.

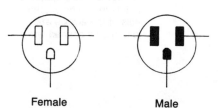

Female
(receptacle)   Male
(plug)

4. Convenience outlet and plug (line-voltage power connectors.) Typical three-prong type. Two-prong and four-prong similar.

GRAPHIC SYMBOLS (cont'd.)

CRYSTAL UNIT, PIEZOELECTRIC (Y)

1. Piezoelectric crystal unit, including quartz crystal.

FUSE (F)

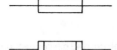

1. General: all three symbols are used.

GROUND, CIRCUIT RETURN   (no class letter)

1. General: either earth, body of water, or chassis at zero potential

2. Chassis ground: may be at substantial potential with respect to earth ground.

INDUCTOR (L)

1. General: right-hand symbol is deprecated and should not be used on new schematics.

2. Magnetic core inductor

GRAPHIC SYMBOLS (cont'd.)

3. Tapped inductor

4. Adjustable inductor

INTEGRATED CIRCUIT (U)

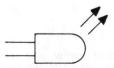

1. General: unused pin connections need not be shown.

The asterisk is not part of the symbol. It indicates where the type number is placed.

LAMP (DS)

1. General; light source, general

2. Glow lamp, neon lamp (AC type)

3. Incandescent lamp

4. Indicating, pilot, signaling or switchboard light

GRAPHIC SYMBOLS (cont'd.)

METER (M)

1. The asterisk is not part of the
   symbol. It indicates where to place a
   letter or letters indicating the type
   of meter:

A = ammeter W = wattmeter
DB, VU = audio level meter
F = frequency meter
MA = milliammeter
OHM = ohmmeter V = voltmeter

MICROPHONE (MK)   Microphone

Microphone

RELAY (K)

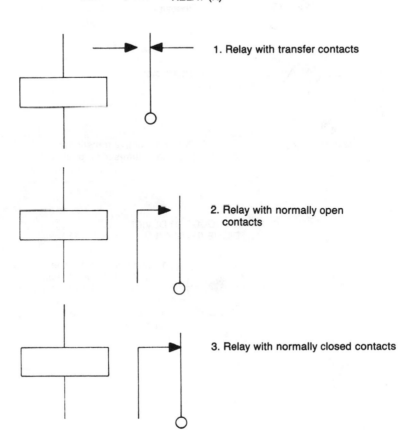

1. Relay with transfer contacts

2. Relay with normally open
   contacts

3. Relay with normally closed contacts

GRAPHIC SYMBOLS (cont'd.)

RESISTOR (R)

1. General

2. Tapped resistor

3. Resistor with adjustable contact
   (potentiometer)

4. Continuously adjustable resistor
   (rheostat)

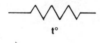

5. Thermistor

6. Photoconductive transducer
   (e.g., cadmium-sulfide photo-cell)

SEMICONDUCTOR DEVICE
DIODE (D OR CR)

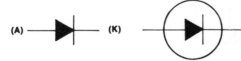

1. Semiconductor diode; enclosure
   symbol may be omitted where
   confusion would not be caused.

   A = anode, K = cathode. The
   letters are not part of the symbol.

2. Breakdown diode (zener diode)

GRAPHIC SYMBOLS (cont'd.)

3. Tunnel diode

4. Photosensitive diode

5. Photoemissive diode, light-emitting diode (LED)

6. Thyristor or silicon controlled rectifier (SCR)

7. Triac

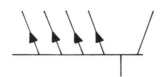

8. Diac (same as triac, but no gate lead)

TRANSISTOR (Q)

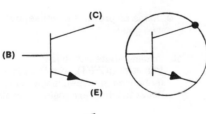

(C)

(B)

(E)

1. NPN transistor; enclosure symbol may be omitted where confusion would not be caused (unless an electrode is connected to it, as shown here).

C = collector, E = emitter, B = base; these letters are not part of the symbol.

2. PNP transistor

3. NPN transistor with multiple emitters (four shown in this example)

GRAPHIC SYMBOLS (cont'd.)

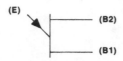

4. Unijunction transistor with N-type base. If arrow on emitter points in opposite direction base is P type.

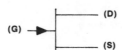

5. Junction field-effect transistor (JFET) with N-channel junction gate.

G = gate, D = drain, S = source; these letters are not part of the symbol.

6. Insulated-gate field-effect transistor (IGFET) with N-channel (depletion type), single gate, positive substrate.

7. Insulated-gate field-effect transistor (IGFET), with N-channel (depletion type), single gate, active substrate internally terminated to source.

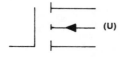

8. Insulated-gate field-effect transistor (IGFET) with N-channel (enhancement type), single gate, active substrate externally terminated.

U = substrate; this letter is not part of symbol.

9. Same as previous example, but with two gates

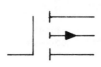

10. Insulated-gate field-effect transistor (IGFET), with P-channel (enhancement type), single gate, active substrate externally terminated

11. Phototransistor (NPN type)

12. Photovoltaic transducer; barrier photocell; solar cell (No class designation letter)

GRAPHIC SYMBOLS (cont'd.)

SWITCH (S)

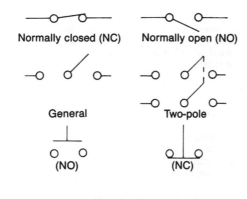

Normally closed (NC)　　Normally open (NO)

General　　　　Two-pole

(NO)　　　　(NC)

1. Single-throw switch; terminals are necessary for clarity in an NC switch, but may be omitted in an NO switch.

2. Double-throw switch

3. Push button

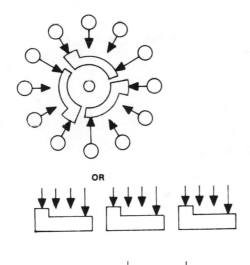

Break-before-make

Make-before-break

4. Selector or multiposition; any number of transmission paths may be shown

OR

5. Rotary or wafer-type switch. Viewed from end opposite control knob. For more than one section the first is the one nearest the control knob. With contacts on both sides front contacts are nearest control knob.

6. Flasher; self-interrupting switch

GRAPHIC SYMBOLS (cont'd.)

SYNCHRO (B)

1. General: if identification is required, add appropriate letter combination from following list adjacent to symbol:

CDX control-differential transmitter
CT   control transformer
CX   control transmitter
TDR torque-differential receiver
TDX torque-differential transmitter
TR   torque receiver
TX   torque transmitter
RS   resolver

2. Synchro: control transformer; receiver; transmitter

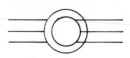

3. Synchro: differential receiver; differential transmitter

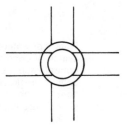

4. Synchro: resolver

PICKUP HEAD (PU)

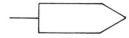

1. General

2. Writing or recording head

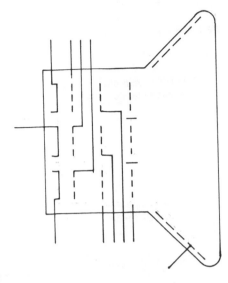

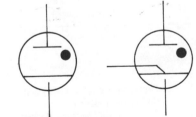

Table of Number Functions (cont'd.)

Number	$\sqrt[3]{Number}$	$\sqrt[3]{10 \times Number}$	$\sqrt[3]{100 \times Number}$
25	2.924018	6.299605	13.57209
26	2.962496	6.382504	13.75069
27	3.000000	6.463304	13.92477
28	3.036589	6.542133	14.09460
29	3.072317	6.619106	14.26043
30	3.107233	6.694330	14.42250
31	3.141381	6.767899	14.58100
32	3.174802	6.839904	14.73613
33	3.207534	6.910423	14.88806
34	3.239612	6.979532	15.03695
35	3.271066	7.047299	15.18294
36	3.301927	7.113787	15.32619
37	3.332222	7.179054	15.46680
38	3.361975	7.243156	15.60491
39	3.391211	7.306144	15.74061
40	3.419952	7.368063	15.87401
41	3.448217	7.428959	16.00521
42	3.476027	7.488872	16.13429
43	3.503398	7.547842	16.26133
44	3.530348	7.605905	16.38643
45	3.556893	7.663094	16.50964
46	3.583048	7.719443	16.63103
47	3.608826	7.774980	16.75069
48	3.634241	7.829735	16.86865
49	3.659306	7.883735	16.98499
50	3.684031	7.937005	17.09976
51	3.708430	7.989570	17.21301
52	3.732511	8.041452	17.32478
53	3.756286	8.092672	17.43513
54	3.779763	8.143253	17.54411
55	3.802952	8.193213	17.65174
56	3.825862	8.242571	17.75808
57	3.848501	8.291344	17.86316
58	3.870877	8.339551	17.96702
59	3.892996	8.387207	18.06969
60	3.914868	8.434327	18.17121
61	3.936497	8.480926	18.27160
62	3.957892	8.527019	18.37091
63	3.979057	8.572619	18.46915
64	4.000000	8.617739	18.56636
65	4.020726	8.662391	18.66256

John D. Lenk, *Practical Semiconductor Data Book for Electronic Engineers and Technicians,* ©1970. Reprinted by permission of Prentice-Hall, Inc.

GRAPHIC SYMBOLS (cont'd.)

 3. Playback or reading head

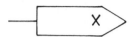 4. Erase head

 5. Stereo head

PIEZOELECTRIC CRYSTAL UNIT (Y)

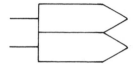 1. Also called quartz crystal unit

TRANSFORMER (T)

  1. General: International symbol on right

 2. Magnetic core, non-saturating

 3. Shielded transformer with magnetic core. A ferrite core is often shown by dashed lines, with arrow if tunable.

GRAPHIC SYMBOLS (cont'd.)

4. Magnetic core with electrostatic shield between windings. (Shield shown connected to frame.)

5. Saturating transformer

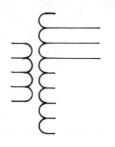

6. Transformer with taps

7. Autotransformer

Fixed   Adjustable

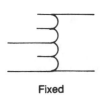

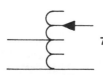

GRAPHIC SYMBOLS (cont'd.)

VACUUM TUBE (V)

1. Triode with directly-heated filamentary cathode

2. Triode with indirectly-heated cathode (heater included)

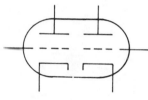

3. Twin triode with indirectly-heated cathode (heater omitted)

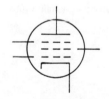

4. Pentode with indirectly-heated cathode (heater omitted)

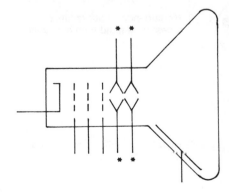

5. Cathode-ray tube (CRT) with deflection plates (*).
Same symbol without deflection plates is used for monochrome picture tube (single electron gun, magnetic deflection).

GRAPHIC SYMBOLS (cont'd.)

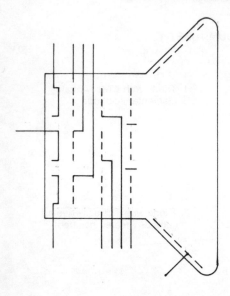

6. Color picture tube with three electron guns and electromagnetic deflection

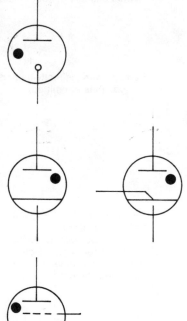

7. Cold-cathode gas-filled tube

8. Mercury-pool vapor-rectifier tubes; right-hand tube has ignitor.

9. Thyratron with indirectly-heated cathode (heater shown)

GREEK ALPHABET

Name	Capital	Small	Designates
Alpha	A	α	Angles, coefficients, attenuation constant, absorption factor, area
Beta	B	β	Angles, coefficients, phase constant
Gamma	Γ	γ	Complex propagation constant (CAP), specific gravity, angles, electrical conductivity, propagation constant
Delta	Δ	δ	Increment or decrement*, determinant (CAP), permittivity (CAP), density, angles
Epsilon	E	ϵ	Dielectric constant, permittivity, base of natural logarithms, electric intensity
Zeta	Z	ζ	Coordinates, coefficients
Eta	H	η	Intrinsic impedance, efficiency, surface charge density, hysteresis, coordinates
Theta	Θ	θ	Angular phase displacement, time constant, reluctance, angles
Iota	I	ι	Unit vector
Kappa	K	κ	Susceptibility, coupling coefficient
Lambda	Λ	λ	Permeance (CAP), wavelength, attenuation constant
Mu	M	μ	Permeability, amplification factor, prefix micro-
Nu	N	ν	Reluctivity, frequency
Xi	Ξ	ξ	Coordinates
Omicron	O	o	
Pi	Π	π	3.141 592 653 589 793 238 ...
Rho	P	ρ	Resistivity, volume charge density, coordinates.
Sigma	Σ	σ	Summation (CAP), surface charge density, complex propagation constant, electrical conductivity, leakage coefficient, deviation
Tau	T	τ	Time constant, volume resistivity, time-phase displacement, transmission factor, density
Upsilon	Υ	υ	
Phi	Φ	ϕ	Scalar potential (CAP), magnetic flux, angles
Chi	X	χ	Electric susceptibility, angles
Psi	Ψ	ψ	Dielectric flux, phase difference, coordinates, angles
Omega	Ω	ω	Ohms (CAP), solid angle (CAP), angular velocity

All use small letter unless capital (CAP) is indicated. An asterisk (*) indicates either capital or small may be used.

MATHEMATICAL TABLES

Common Logarithms

N	0	1	2	3	4	5	6	7	8	9	N
10	0000	0043	0086	0128	0170	0212	0253	0294	0334	0374	10
11	0414	0453	0492	0531	0569	0607	0645	0682	0719	0755	11
12	0792	0828	0864	0899	0934	0969	1004	1038	1072	1106	12
13	1139	1173	1206	1239	1271	1303	1335	1367	1399	1430	13
14	1461	1492	1523	1553	1584	1614	1644	1673	1703	1732	14
15	1761	1790	1818	1847	1875	1903	1931	1959	1987	2014	15
16	2041	2068	2095	2122	2148	2175	2201	2227	2253	2279	16
17	2304	2330	2355	2380	2405	2430	2455	2480	2504	2529	17
18	2553	2577	2601	2625	2648	2672	2695	2718	2742	2765	18
19	2788	2810	2833	2856	2878	2900	2923	2945	2967	2989	19
20	3010	3032	3054	3075	3096	3118	3139	3160	3181	3201	20
21	3222	3243	3263	3284	3304	3324	3345	3365	3385	3404	21
22	3424	3444	3464	3483	3502	3522	3541	3560	3579	3598	22
23	3617	3636	3655	3674	3692	3711	3729	3747	3766	3784	23
24	3802	3820	3838	3856	3874	3892	3909	3927	3945	3962	24
25	3979	3997	4014	4031	4048	4065	4082	4099	4116	4133	25
26	4150	4166	4183	4200	4216	4232	4249	4265	4281	4298	26
27	4314	4330	4346	4362	4378	4393	4409	4425	4440	4456	27
28	4472	4487	4502	4518	4533	4548	4564	4579	4594	4609	28
29	4624	4639	4654	4669	4683	4698	4713	4728	4742	4757	29
30	4771	4786	4800	4814	4829	4843	4857	4871	4886	4900	30
31	4914	4928	4942	4955	4969	4983	4997	5011	5024	5038	31
32	5051	5065	5079	5092	5105	5119	5132	5145	5159	5172	32
33	5185	5198	5211	5224	5237	5250	5263	5276	5289	5302	33
34	5315	5328	5340	5353	5366	5378	5391	5403	5416	5428	34
35	5441	5453	5465	5478	5490	5502	5514	5527	5539	5551	35
36	5563	5575	5587	5599	5611	5623	5635	5647	5658	5670	36
37	5682	5694	5705	5717	5729	5740	5752	5763	5775	5786	37
38	5798	5809	5821	5832	5843	5855	5866	5877	5888	5899	38
39	5911	5922	5933	5944	5955	5966	5977	5988	5999	6010	39
40	6021	6031	6042	6053	6064	6075	6085	6096	6107	6117	40
41	6128	6138	6149	6160	6170	6180	6191	6201	6212	6222	41
42	6232	6243	6253	6263	6274	6284	6294	6304	6314	6325	42
43	6335	6345	6355	6365	6375	6385	6395	6405	6415	6425	43
44	6435	6444	6454	6464	6474	6484	6493	6503	6513	6522	44
45	6532	6542	6551	6561	6571	6580	6590	6599	6609	6618	45
46	6628	6637	6646	6656	6665	6675	6684	6693	6702	6712	46
47	6721	6730	6739	6749	6758	6767	6776	6785	6794	6803	47
48	6812	6821	6830	6839	6848	6857	6866	6875	6884	6893	48
49	6902	6911	6920	6928	6937	6946	6955	6964	6972	6981	49
50	6990	6998	7007	7016	7024	7033	7042	7050	7059	7067	50
51	7076	7084	7093	7101	7110	7118	7126	7135	7143	7152	51
52	7160	7168	7177	7185	7193	7202	7210	7218	7226	7235	52
53	7243	7251	7259	7267	7275	7284	7292	7300	7308	7316	53
54	7324	7332	7340	7348	7356	7364	7372	7380	7388	7396	54
N	0	1	2	3	4	5	6	7	8	9	N

Common Logarithms (Continued)

N	0	1	2	3	4	5	6	7	8	9	N
55	7404	7412	7419	7427	7435	7443	7451	7459	7466	7474	55
56	7482	7490	7497	7505	7513	7520	7528	7536	7543	7551	56
57	7559	7566	7574	7582	7589	7597	7604	7612	7619	7627	57
58	7634	7642	7649	7657	7664	7672	7679	7686	7694	7701	58
59	7709	7716	7723	7731	7738	7745	7752	7760	7767	7774	59
60	7782	7789	7796	7803	7810	7818	7825	7832	7839	7846	60
61	7853	7860	7868	7875	7882	7889	7896	7903	7910	7917	61
62	7924	7931	7938	7945	7952	7959	7966	7973	7980	7987	62
63	7993	8000	8007	8014	8021	8028	8035	8041	8048	8055	63
64	8062	8069	8075	8082	8089	8096	8102	8109	8116	8122	64
65	8129	8136	8142	8149	8156	8162	8169	8176	8182	8189	65
66	8195	8202	8209	8215	8222	8228	8235	8241	8248	8254	66
67	8261	8267	8274	8280	8287	8293	8299	8306	8312	8319	67
68	8325	8331	8338	8344	8351	8357	8363	8370	8376	8382	68
69	8388	8395	8401	8407	8414	8420	8426	8432	8439	8445	69
70	8451	8457	8463	8470	8476	8482	8488	8494	8500	8506	70
71	8513	8519	8525	8531	8537	8543	8549	8555	8561	8567	71
72	8573	8579	8585	8591	8597	8603	8609	8615	8621	8627	72
73	8633	8639	8645	8651	8657	8663	8669	8675	8681	8686	73
74	8692	8698	8704	8710	8716	8722	8727	8733	8739	8745	74
75	8751	8756	8762	8768	8774	8779	8785	8791	8797	8802	75
76	8808	8814	8820	8825	8831	8837	8842	8848	8854	8859	76
77	8865	8871	8876	8882	8887	8893	8899	8904	8910	8915	77
78	8921	8927	8932	8938	8943	8949	8954	8960	8965	8971	78
79	8976	8982	8987	8993	8998	9004	9009	9015	9020	9025	79
80	9031	9036	9042	9047	9053	9058	9063	9069	9074	9079	80
81	9085	9090	9096	9101	9106	9112	9117	9122	9128	9133	81
82	9138	9143	9149	9154	9159	9165	9170	9175	9180	9186	82
83	9191	9196	9201	9206	9212	9217	9222	9227	9232	9238	83
84	9243	9248	9253	9258	9263	9269	9274	9279	9284	9289	84
85	9294	9299	9304	9309	9315	9320	9325	9330	9335	9340	85
86	9345	9350	9355	9360	9365	9370	9375	9380	9385	9390	86
87	9395	9400	9405	9410	9415	9420	9425	9430	9435	9440	87
88	9445	9450	9455	9460	9465	9469	9474	9479	9484	9489	88
89	9494	9499	9504	9509	9513	9518	9523	9528	9533	9538	89
90	9542	9547	9552	9557	9562	9566	9571	9576	9581	9586	90
91	9590	9595	9600	9605	9609	9614	9619	9624	9628	9633	91
92	9638	9643	9647	9652	9657	9661	9666	9671	9675	9680	92
93	9685	9689	9694	9699	9703	9708	9713	9717	9722	9727	93
94	9731	9736	9741	9745	9750	9754	9759	9763	9768	9773	94
95	9777	9782	9786	9791	9795	9800	9805	9809	9814	9818	95
96	9823	9827	9832	9836	9841	9845	9850	9854	9859	9863	96
97	9868	9872	9877	9881	9886	9890	9894	9899	9903	9908	97
98	9912	9917	9921	9926	9930	9934	9939	9943	9948	9952	98
99	9956	9961	9965	9969	9974	9978	9983	9987	9991	9996	99
N	0	1	2	3	4	5	6	7	8	9	N

Table of Sines, Cosines, and Tangents

Angle	Radians	Sine	Cosine	Tangent	Angle	Radians	Sine	Cosine	Tangent
0°	.0000	.0000	1.0000	.0000	45°	.7854	.7071	.7071	1.0000
1	.0175	.0175	.9998	.0175	46	.8029	.7193	.6947	1.0355
2	.0349	.0349	.9994	.0349	47	.8203	.7314	.6820	1.0724
3	.0524	.0523	.9986	.0524	48	.8378	.7431	.6691	1.1106
4	.0698	.0698	.9976	.0699	49	.8552	.7547	.6561	1.1504
5	.0873	.0872	.9962	.0875	50	.8727	.7660	.6428	1.1918
6	.1047	.1045	.9945	.1051	51	.8901	.7771	.6293	1.2349
7	.1222	.1219	.9925	.1228	52	.9076	.7880	.6157	1.2799
8	.1396	.1392	.9903	.1405	53	.9250	.7986	.6018	1.3270
9	.1571	.1564	.9877	.1584	54	.9425	.8090	.5878	1.3764
10	.1745	.1736	.9848	.1763	55	.9599	.8192	.5736	1.4281
11	.1920	.1908	.9816	.1944	56	.9774	.8290	.5592	1.4826
12	.2094	.2079	.9781	.2126	57	.9948	.8387	.5446	1.5399
13	.2269	.2250	.9744	.2309	58	1.0123	.8480	.5299	1.6003
14	.2443	.2419	.9703	.2493	59	1.0297	.8572	.5150	1.6643
15	.2618	.2588	.9659	.2679	60	1.0472	.8660	.5000	1.7321
16	.2793	.2756	.9613	.2867	61	1.0647	.8746	.4848	1.8040
17	.2967	.2924	.9563	.3057	62	1.0821	.8829	.4695	1.8807
18	.3142	.3090	.9511	.3249	63	1.0996	.8910	.4540	1.9626
19	.3316	.3256	.9455	.3443	64	1.1170	.8988	.4384	2.0503
20	.3491	.3420	.9397	.3640	65	1.1345	.9063	.4226	2.1445
21	.3665	.3584	.9336	.3839	66	1.1519	.9135	.4067	2.2460
22	.3840	.3746	.9272	.4040	67	1.1694	.9205	.3907	2.3559
23	.4014	.3907	.9205	.4245	68	1.1868	.9272	.3746	2.4751
24	.4189	.4067	.9135	.4452	69	1.2043	.9336	.3584	2.6051
25	.4363	.4226	.9063	.4663	70	1.2217	.9397	.3420	2.7475
26	.4538	.4384	.8988	.4877	71	1.2392	.9455	.3256	2.9042
27	.4712	.4540	.8910	.5095	72	1.2566	.9511	.3090	3.0777
28	.4887	.4695	.8829	.5317	73	1.2741	.9563	.2924	3.2709
29	.5061	.4848	.8746	.5543	74	1.2915	.9613	.2756	3.4874
30	.5236	.5000	.8660	.5774	75	1.3090	.9659	.2588	3.7321
31	.5411	.5150	.8572	.6009	76	1.3265	.9703	.2419	4.0108
32	.5585	.5299	.8480	.6249	77	1.3439	.9744	.2250	4.3315
33	.5760	.5446	.8387	.6494	78	1.3614	.9781	.2079	4.7046
34	.5934	.5592	.8290	.6745	79	1.3788	.9816	.1908	5.1446
35	.6109	.5736	.8192	.7002	80	1.3963	.9848	.1736	5.6713
36	.6283	.5878	.8090	.7265	81	1.4137	.9877	.1564	6.3138
37	.6458	.6018	.7986	.7536	82	1.4312	.9903	.1392	7.1154
38	.6632	.6157	.7880	.7813	83	1.4486	.9925	.1219	8.1443
39	.6807	.6293	.7771	.8098	84	1.4661	.9945	.1045	9.5144
40	.6981	.6428	.7660	.8391	85	1.4835	.9962	.0872	11.43
41	.7156	.6561	.7547	.8693	86	1.5010	.9976	.0698	14.30
42	.7330	.6691	.7431	.9004	87	1.5184	.9986	.0523	19.08
43	.7505	.6820	.7314	.9325	88	1.5359	.9994	.0349	28.64
44	.7679	.6947	.7193	.9657	89	1.5533	.9998	.0175	57.29

Table of Number Functions

Number	Number²	√Number	√10 × Number	Number³
1	1	1.000000	3.162278	1
2	4	1.414214	4.472136	8
3	9	1.732051	5.477226	27
4	16	2.000000	6.324555	64
5	25	2.236068	7.071068	125
6	36	2.449490	7.745967	216
7	49	2.645751	8.366600	343
8	64	2.828427	8.944272	512
9	81	3.000000	9.486833	729
10	100	3.162278	10.00000	1,000
11	121	3.316625	10.48809	1,331
12	144	3.464102	10.95445	1,728
13	169	3.605551	11.40175	2,197
14	196	3.741657	11.83216	2,744
15	225	3.872983	12.24745	3,375
16	256	4.000000	12.64911	4,096
17	289	4.123106	13.03840	4,913
18	324	4.242641	13.41641	5,832
19	361	4.358899	13.78405	6,859
20	400	4.472136	14.14214	8,000
21	441	4.582576	14.49138	9,261
22	484	4.690416	14.83240	10,648
23	529	4.795832	15.16575	12,167
24	576	4.898979	15.49193	13,824
25	625	5.000000	15.81139	15,625
26	676	5.099020	16.12452	17,576
27	729	5,196152	16.43168	19,683
28	784	5.291503	16.73320	21,952
29	841	5.385165	17.02939	24,389
30	900	5.477226	17.32051	27,000
31	961	5.567764	17.60682	29,791
32	1,024	5.656854	17.88854	32,768
33	1,089	5.744563	18.16590	35,937
34	1,156	5.830952	18.43909	39,304
35	1,225	5.916080	18.70829	42,875
36	1,296	6.000000	18.97367	46,656
37	1,369	6.082763	19.23538	50,653
38	1,444	6.164414	19.49359	54,872
39	1,521	6.244998	19.74842	59,319
40	1,600	6.324555	20.00000	64,000
41	1,681	6.403124	20.24846	68,921
42	1,764	6.480741	20.49390	74,088
43	1,849	6.557439	20.73644	79,507
44	1,936	6.633250	20.97618	85,184

John D. Lenk, *Practical Semiconductor Data Book for Electronic Engineers and Technicians,* ©1970. Reprinted by permission of Prentice-Hall, Inc.

Table of Number Functions (cont'd.)

Number	Number^2	$\sqrt{Number}$	$\sqrt{10 \times Number}$	Number^3
45	2,025	6.708204	21.21320	91,125
46	2,116	6.782330	21.44761	97,336
47	2,209	6.855655	21.67948	103,823
48	2,304	6.928203	21.90890	110,592
49	2,401	7.000000	22.13594	117,649
50	2,500	7.071680	22.36068	125,000
51	2,601	7.141428	22.58318	132,651
52	2,704	7.211103	22.80351	140,608
53	2,809	7.280110	23.02173	148,877
54	2,916	7.348469	23.23790	157,464
55	3,025	7.416198	23.45208	166,375
56	3,136	7.483315	23.66432	175,616
57	3,249	7.549834	23.87467	185,193
58	3,364	7.615773	24.06319	194,112
59	3,481	7.681146	24.28992	205,379
60	3,600	7.745967	24.49490	216,000
61	3,721	7.810250	24.69818	226,981
62	3,844	7.874008	24.89980	238,047
63	3,969	7.937254	25.09980	250,047
64	4,096	8.000000	25.29822	262,144
65	4,225	8.062258	25.49510	274,625
66	4,356	8.124038	25.69047	287,496
67	4,489	8.185353	25.88436	300,763
68	4,624	8.246211	26.07681	314,432
69	4,761	8.306624	26.26785	328,509
70	4,900	8.366600	26.45751	343,000
71	5,041	8.426150	26.64583	357,911
72	5,184	8.485281	26.83282	373,248
73	5,329	8.544004	27.01851	389,017
74	5,476	8.602325	27.20294	405,224
75	5,625	8.660254	27.38613	421,875
76	5,776	8.717798	27.56810	438,976
77	5,929	8.774964	27.74887	456,533
78	6,084	8.831761	27.92848	474,552
79	6,241	8.888194	28.10694	493,039
80	6,400	8.944272	28.28427	512,000
81	6,561	9.000000	28.46050	531,441
82	6,724	9.055385	28.63564	551,368
83	6,889	9.110434	28.80972	571,787
84	7,056	9.165151	28.98275	592,704
85	7,225	9.219544	29.15476	614,125

John D. Lenk, *Practical Semiconductor Data Book for Electronic Engineers and Technicians,* ©1970. Reprinted by permission of Prentice-Hall, Inc.

Table of Number Functions (cont'd.)

Number	*Number*²	$\sqrt{Number}$	$\sqrt{10 \times Number}$	*Number*³
86	7,396	9.273618	29.32576	636,056
87	7,569	9.327379	29.49576	658,503
88	7,744	9.380832	29.66479	681,472
89	7,921	9.433981	29.83287	704,969
90	8,100	9.486833	30.00000	729,000
91	8,281	9.539392	30.16621	753,571
92	8,464	9.591663	30.33150	778,688
93	8,649	9.643651	30.49590	804,357
94	8,836	9.695360	30.65942	830,584
95	9,025	9.746794	30.82207	857,375
96	9,216	9.797959	30.98387	884,736
97	9,409	9.848858	31.14482	912,673
98	9,604	9.899495	31.30495	941,192
99	9,801	9.949874	31.46427	970,299
100	10,000	10.00000	31.62278	1,000,000

Number	$\sqrt[3]{Number}$	$\sqrt[3]{10 \times Number}$	$\sqrt[3]{100 \times Number}$
1	1.000000	2.154435	4.641589
2	1.259921	2.714418	5.848035
3	1.442250	3.107233	6.694330
4	1.587401	3.419952	7.368063
5	1.709976	3.684031	7.937005
6	1.817121	3.914868	8.434327
7	1.912931	4.121285	8.879040
8	2.000000	4.308869	9.283178
9	2.080084	4.481405	9.654894
10	2.154435	4.641589	10.00000
11	2.223980	4.791420	10.32280
12	2.289428	4.932424	10.62659
13	2.351335	5.065797	10.91393
14	2.410142	5.192494	11.18689
15	2.466212	5.313293	11.44714
16	2.519842	5.428835	11.69607
17	2.571282	5.539658	11.93483
18	2.620741	5.646216	12.16440
19	2.668402	5.748897	12.38562
20	2.714418	5.848035	12.59921
21	2.758924	5.943922	12.80579
22	2.802039	6.036811	13.00591
23	2.843867	6.126926	15.20006
24	2.884499	6.214465	13.38866

John D. Lenk, *Practical Semiconductor Data Book for Electronic Engineers and Technicians,* ©1970. Reprinted by permission of Prentice-Hall, Inc.

Table of Number Functions (cont'd.)

Number	$\sqrt[3]{Number}$	$\sqrt[3]{10 \times Number}$	$\sqrt[3]{100 \times Number}$
25	2.924018	6.299605	13.57209
26	2.962496	6.382504	13.75069
27	3.000000	6.463304	13.92477
28	3.036589	6.542133	14.09460
29	3.072317	6.619106	14.26043
30	3.107233	6.694330	14.42250
31	3.141381	6.767899	14.58100
32	3.174802	6.839904	14.73613
33	3.207534	6.910423	14.88806
34	3.239612	6.979532	15.03695
35	3.271066	7.047299	15.18294
36	3.301927	7.113787	15.32619
37	3.332222	7.179054	15.46680
38	3.361975	7.243156	15.60491
39	3.391211	7.306144	15.74061
40	3.419952	7.368063	15.87401
41	3.448217	7.428959	16.00521
42	3.476027	7.488872	16.13429
43	3.503398	7.547842	16.26133
44	3.530348	7.605905	16.38643
45	3.556893	7.663094	16.50964
46	3.583048	7.719443	16.63103
47	3.608826	7.774980	16.75069
48	3.634241	7.829735	16.86865
49	3.659306	7.883735	16.98499
50	3.684031	7.937005	17.09976
51	3.708430	7.989570	17.21301
52	3.732511	8.041452	17.32478
53	3.756286	8.092672	17.43513
54	3.779763	8.143253	17.54411
55	3.802952	8.193213	17.65174
56	3.825862	8.242571	17.75808
57	3.848501	8.291344	17.86316
58	3.870877	8.339551	17.96702
59	3.892996	8.387207	18.06969
60	3.914868	8.434327	18.17121
61	3.936497	8.480926	18.27160
62	3.957892	8.527019	18.37091
63	3.979057	8.572619	18.46915
64	4.000000	8.617739	18.56636
65	4.020726	8.662391	18.66256

Table of Number Functions (cont'd.)

Number	$\sqrt[3]{Number}$	$\sqrt[3]{10 \times Number}$	$\sqrt[3]{100 \times Number}$
66	4.041240	8.706588	18.75777
67	4.061548	8.750340	18.85204
68	4.081655	8.793659	18.94536
69	4.101566	8.836556	19.03778
70	4.121285	8.879040	19.12931
71	4.140818	8.921121	19.21997
72	4.160168	8.962809	19.30979
73	4.179339	9.004113	19.39877
74	4.198336	9.045042	19.48695
75	4.217163	9.085603	19.57434
76	4.235824	9.125805	19.66095
77	4.254321	9.165656	19.74681
78	4.272659	9.205164	19.83192
79	4.290840	9.244335	19.91632
80	4.308869	9.283178	20.00000
81	4.326749	9.321698	20.08299
82	4.344481	9.359902	20.16530
83	4.362071	9.397796	20.24694
84	4.379519	9.435388	20.32793
85	4.396830	9.472682	20.40828
86	4.414005	9.509685	20.48800
87	4.431048	9.546403	20.56710
88	4.447960	9.582840	20.64560
89	4.464745	9.619002	20.72351
90	4.481405	9.654894	20.80084
91	4.497941	9.690521	20.87759
92	4.514357	9.725888	20.95379
93	4.530655	9.761000	21.02944
94	4.546836	9.795861	21.10454
95	4.562903	9.830476	21.17912
96	4.578857	9.864848	21.25317
97	4.594701	9.898983	21.32671
98	4.610436	9.932884	21.39975
99	4.626065	9.966555	21.47229
100	4.641589	10.00000	21.54435

John D. Lenk, *Practical Semiconductor Data Book for Electronic Engineers and Technicians,* ©1970. Reprinted by permission of Prentice-Hall, Inc.

PREFERRED VALUES

In order to limit the quantities of parts that must be stocked and to standardize their values, preferred numbers are used. These are calculated according to their tolerances, each nominal value being separated from the next by a constant multiplier. For small electronic components, such as fixed composition resistors and fixed ceramic, mica and molded capacitors, the following values are used:

TABLE I

ANSI Standard C83.2-1971

±20%	±10%	±5%
10	10	10
		11
	12	12
		13
15	15	15
		16
	18	18
		20
22	22	22
		24
	27	27
		30
33	33	33
		36
	39	39
		43
47	47	47
		51
	56	56
		62
68	68	68
		75
	82	82
		91
100	100	100

A slightly different set of preferred values is used for fixed wire-wound power-type resistors and for time-delay fuses:

TABLE II
ANSI STANDARD Z17.1-1973
Series "5" (±24%)   Series "10" (±12%)

Series "5" (±24%)	Series "10" (±12%)
10	10
	12
16	16
	20
25	25
	32
40	40
	50
63	63
	80
100	100

These tables give the two significant figures of each value, which could therefore be, for example, 33, 330, 3300, 33,000, 330,000, or 3,300,000, and so on. Those numbers that appear in more than one column are values available in more than one tolerance, the tolerance of each column in which they appear.

Not all manufacturers adhere to the preferred values, especially in the area of wire-wound resistors, so catalogs should be consulted. "Mil-spec" resistors, which are often available from surplus outlets, have the following values if their tolerance is ±1 percent (multiply by multiples of 10 for higher values):

Ohms	Ohms	Ohms	Ohms	Ohms	Ohms	Ohms	Ohms
1.00	1.33	1.78	2.37	3.16	4.22	5.62	7.50
1.02	1.37	1.82	2.43	3.24	4.32	5.76	7.68
1.05	1.40	1.87	2.49	3.32	4.42	5.90	7.87
1.07	1.43	1.91	2.55	3.40	4.53	6.04	8.06
1.10	1.47	1.96	2.61	3.48	4.64	6.19	8.25
1.13	1.50	2.00	2.67	3.57	4.75	6.34	8.45
1.15	1.54	2.05	2.74	3.65	4.87	6.49	8.66
1.18	1.58	2.10	2.80	3.74	4.99	6.65	8.87
1.21	1.62	2.15	2.87	3.83	5.11	6.81	9.09
1.24	1.65	2.21	2.94	3.92	5.23	6.98	9.31
1.27	1.69	2.26	3.01	4.02	5.36	7.15	9.53
1.30	1.74	2.32	3.09	4.12	5.49	7.32	9.76

PROPERTIES OF MATERIALS
Metals Commonly Used in Electronics

Metal	Density at 20° C g/cm³	Melting Point ° C	Coefficient of Linear Expansion at 20° C × 10⁻⁶/° C	Resistivity μΩ/cm	Modulus of Elasticity kg/mm³	Thermal Conductivity at 20° C W/cm/° C/s
Aluminum	2.70	660	22.90	2.62	7 250	2.18
Beryllium	1.82	1 278	12.00	10.00	30 000	1.64
Brass	8.55	900	18.77	3.90	13 200	*
Bronze	8.15	1 040	18.45	6.50	16 500	*
Copper	8.96	1 083	16.50	1.67	11 000	3.94
Gold	19.30	1 063	14.20	2.19	7 300	2.96
Iridium	22.40	2 410	6.50	5.30	52 500	1.40
Iron (wrought)	7.87	1 535	11.70	9.71	20 000	0.79
Lead	11.34	327	28.70	21.90	1 800	0.35
Magnesium	1.74	651	25.20	4.46	4 600	1.55
Manganese	7.44	1 244	23.00	5.00	16 000	
Mercury	13.55	−39		95.80		0.08
Molybdenum	10.20	2 610	4.90	4.90	35 000	1.46
Monel	8.90	1 400		42.00		
Nickel	8.90	1 453	13.30	6.84	21 000	0.90
Osmium	22.48	3 000	5.00	9.50	57 000	0.61
Palladium	12.00	1 552	11.80	10.80	12 000	0.70
Platinum	21.45	1 769	8.90	9.83	15 000	0.69
Rhodium	12.00	1 966	8.10	4.51	30 000	1.50
Ruthenium	12.20	2 250	9.10	7.60	42 000	
Silver	10.49	961	18.90	1.59	7 200	4.08
Tantalum	16.60	2 996	6.60	12.40	19 000	0.54
Tin	7.30	232	23.00	11.40	41 100	0.64
Titanium	4.54	1 675	8.50	80.00	8 500	0.20
Tungsten	19.30	3 410	4.30	5.50	35 000	1.99
Zinc	7.14	419	29.39	6.00	8 400	1.10
Zirconium	6.40	1 852	5.60	41.00	7 500	1.40

---

*Varies with composition, somewhat lower than copper

Table of Standard Annealed Bare Copper Wire
Using American Wire Gauge (B & S)

Gauge (AWG) or (B & S)	DIAMETER INCHES			AREA	WEIGHT	LENGTH	RESISTANCE AT 68° F			Current Capacity (Amps)— Rubber Insulated
	Min.	Nom.	Max.	Circular Mils	Pounds per M'	Feet per Lb.	Ohms per M'	Feet per Ohm	Ohms per Lb.	
0000	.4554	.4600	.4646	211600.	640.5	1.561	.04901	20400.	.00007652	225
000	.4055	.4096	.4157	167800.	507.9	1.968	.06180	16180.	.0001217	175
00	.3612	.3648	.3684	133100.	402.8	2.482	.07793	12830.	.0001935	150
0	.3217	.3249	.3281	105500.	319.5	3.130	.09827	10180.	.0003076	125
1	.2864	.2893	.2922	83690.	253.3	3.947	.1239	8070.	.0004891	100
2	.2550	.2576	.2602	66370.	200.9	4.977	.1563	6400.	.0007778	90
3	.2271	.2294	.2317	52640.	159.3	6.276	.1970	5075.	.001237	80
4	.2023	.2043	.2063	41740.	126.4	7.914	.2485	4025.	.001966	70
5	.1801	.1819	.1837	33100.	100.2	9.980	.3133	3192.	.003127	55
6	.1604	.1620	.1636	26250.	79.46	12.58	.3951	2531.	.004972	50
7	.1429	.1443	.1457	20820.	63.02	15.87	.4982	2007.	.007905	
8	.1272	.1285	.1298	16510.	49.98	20.01	.6282	1592.	.01257	35
9	.1133	.1144	.1155	13090.	39.63	25.23	.7921	1262.	.01999	
10	.1009	.1019	.1029	10380.	31.43	31.82	.9989	1001.	.03178	25
11	.08983	.09074	.09165	8234.	24.92	40.12	1.260	794.	.05053	
12	.08000	.08081	.08162	6530.	19.77	50.59	1.588	629.6	.08035	20
13	.07124	.07196	.07268	5178.	15.68	63.80	2.003	499.3	.1278	
14	.06344	.06408	.06472	4107.	12.43	80.44	2.525	396.0	.2032	15
15	.05650	.05707	.05764	3257.	9.858	101.4	3.184	314.0	.3230	
16	.05031	.05082	.05133	2583.	7.818	127.9	4.016	249.0	.5136	6
17	.04481	.04526	.04571	2048.	6.200	161.3	5.064	197.5	.8167	
18	.03990	.04030	.04070	1624.	4.917	203.4	6.385	156.5	1.299	3
19	.03553	.03589	.03625	1288.	3.899	256.5	8.051	124.2	2.065	
20	.03164	.03196	.03228	1022.	3.092	323.4	10.15	98.5	3.283	
21	.02818	.02846	.02874	810.1	2.452	407.8	12.80	78.11	5.221	
22	.02510	.02535	.02560	642.4	1.945	514.2	16.14	61.95	8.301	
23	.02234	.02257	.02280	509.5	1.542	648.4	20.36	49.13	13.20	
24	.01990	.02010	.02030	404.0	1.223	817.7	25.67	38.96	20.99	
25	.01770	.01790	.01810	320.4	.9699	1031.	32.37	30.90	33.37	
26	.01578	.01594	.01610	254.1	.7692	1300.	40.81	24.50	53.06	
27	.01406	.01420	.01434	201.5	.6100	1639.	51.47	19.43	84.37	
28	.01251	.01264	.01277	159.8	.4837	2067.	64.90	15.41	134.2	
29	.01115	.01126	.01137	126.7	.3836	2607.	81.83	12.22	213.3	
30	.00993	.01003	.01013	100.5	.3042	3287.	103.2	9.691	339.2	
31	.008828	.008928	.009028	79.7	.2413	4145.	130.1	7.685	539.3	
32	.007850	.007950	.008050	63.21	.1913	5227.	164.1	6.095	857.6	
33	.006980	.007080	.007180	50.13	.1517	6591.	206.9	4.833	1364.	
34	.006205	.006305	.006405	39.75	.1203	8310.	260.9	3.833	2168.	
35	.005515	.005615	.005715	31.52	.09542	10480.	329.0	3.040	3448.	
36	.004900	.005000	.005100	25.00	.07568	13210.	414.8	2.411	5482.	
37	.004353	.004453	.004553	19.83	.06001	16660.	523.1	1.912	8717.	
38	.003865	.003965	.004065	15.72	.04759	21010.	659.6	1.516	13860.	
39	.003431	.003531	.003631	12.47	.03774	26500.	831.8	1.202	22040.	
40	.003045	.003145	.003245	9.888	.02993	33410.	1049.	0.9534	35040.	
41	.00270	.00280	.00290	7.8400	.02373	42140.	1323.	.7559	55750.	
42	.00239	.00249	.00259	6.2001	.01877	53270.	1673.	.5977	89120.	
43	.00212	.00222	.00232	4.9284	.01492	67020.	2104.	.4753	141000.	
44	.00187	.00197	.00207	3.8809	.01175	85100.	2672.	.3743	227380.	
45	.00166	.00176	.00186	3.0976	.00938	106600.	3348.	.2987	356890.	
46	.00147	.00157	.00167	2.4649	.00746	134040.	4207.	.2377	563900.	

*Note: Values from National Electrical Code.

Principal Semiconductors

Semiconductor	Density (g/cm³)	Melting Point (°C)	Coefficient of Linear Expansion (× 10⁻⁶/°C)	Energy Band Gap at 300K (eV)	Electron Mobility (cm³/V.s)		Hole Mobility (cm³/V.s)	
					Light Mass	Heavy Mass	Light Mass	Heavy Mass
AlSb	4.28	1 065		1.60		180-230		420-500
B	2.34	2 075		1.40	1	1		2
C (diamond)	3.51	3 800	1.18	5.30	1 800	1 800		1 600
GaAs	5.32	1 238	5.70	1.43	8 600-11 000	1 000	3 000	426-500
GaSb	5.62	706	6.90	0.70	5 000-40 000	1 000	7 000	700-1 200
GaP	4.13	1 450	5.30	2.25		120-300		420-500
Ge	5.32	937	6.10	0.66	3 900	3 900	14 000	1 860
InAs	5.67	942	5.30	0.33	33 000-40 000		8 000	450-500
InP	4.79	1 062	4.50	1.27	4 800-6 800			150-200
InSb	5.78	530	5.50	0.17	78 000		12 000	750
Se (amorphous)	4.82			2.30	0.005			0.15
Se (hexagonal)	4.79	217	36.9	1.80				1
Si	2.33	1 417	4.20	1.09	1 500	1 500	1 500	480
Te	6.25	432	16.80	0 38	1 100	1 100	10 000	700

Carrier mobilities are at 300 K

Electrical Properties of Commonly Used Insulators

Material	Resistivity (Ω/cm at 25°C)	Dielectric Constant (at 1MHz at 25°C)
Air	—	1.0
Asbestos	—	3.1
Bakelite	10^{11}	4.4
Beeswax	—	2.5
Glass	—	8.3
Gutta-percha	10^{15}	2.5
Mahogany	—	2.3
Nylon	8×10^{14}	3.1
Paper	—	3.0
Phenol (formaldehyde, 50% paper laminate)	—	4.6
Plywood (Douglas fir)	—	1.9
Porcelain	—	5.1
Polyvinylchloride (PVC)	10^{14}	2.9
Mica, ruby	5×10^{13}	5.4
Shellac	—	3.5
Silicon dioxide	$>10^{19}$	3.8
Silicone-rubber	—	3.2
Teflon	10^{17}	2.1
Vaseline	—	2.2
Water (distilled)	10^6	78.2

Properties of Ferromagnetic Materials
Representative Core Materials

Material	Permeability Initial	Maximum	Coercivity (A/m)	Retentivity (T)
Ferroxcube 3 (Mn-Zn-Ferrite)	1.26×10^{-3}	1.88×10^{-3}	7.96×10^{-2}	0.10
Ferroxcube 101 (Ni-Zn-Ferrite)	1.38×10^{-3}		1.43×10^{-1}	0.11
HyMu 80 (Ni 80%, Fe 20%)	2.51×10^{-2}	1.26×10^{-1}	3.98×10^{-2}	
Iron, silicon (transformer) (Fe 96%, Si 4%)	6.28×10^{-4}	8.80×10^{-3}	2.39×10^{-1}	0.70
Mumetal (Ni 77%, Fe 16%, Cu 5%, Cr 2%)	2.51×10^{-2}	1.26×10^{-1}	3.98×10^{-2}	0.60
Permalloy 45 (Fe 55%, Ni 45%)	3.14×10^{-3}	3.14×10^{-2}	2.39×10^{-1}	
Permendur 2V (Fe 49%, Co 49%, V 2%)	1.01×10^{-3}	5.65×10^{-3}	1.59×10^{0}	1.40
Rhometal (Fe 64%, Ni 36%)	1.26×10^{-3}	6.28×10^{-3}	3.98×10^{-1}	0.36
Sendust (high-frequency powder) (Fe 85%, Si 10%, Al 5%)	3.77×10^{-2}	1.51×10^{-1}	3.98×10^{-2}	0.50
Supermalloy (Ni 79%, Fe 16%, Mo 5%)	1.26×10^{-1}	1.26×10^{0}	1.59×10^{-3}	

Properties of Ferromagnetic Materials
Representative Permanent Magnetic Materials

Material	External Energy $(B_d H_d)$	Coercivity (A/m)	Retentivity (T)
Alnico V (Fe 51%, Co 24%, Ni 14%, Al 8%, Cu 3%)	35 810	45 757	1.20
Alnico VI (Fe 48.75%, Co 24%, Ni 15%, Al 8%, Cu 3%, Ti 1.25%)	27 852	59 683	1.00

Material	External Energy (B_dH_d)	Coercivity (A/m)	Retentivity (T)
Carbon steel (Fe 98.5%, C 1%, Mn 0.5%)	1 432	3 820	0.86
Chromium steel (Fe 95.5%, Cr 3.5%, C 1%)	2 308	5 013	0.90
Cobalt steel (Co 36%, Cr 35%, Fe 25.15%, W 3%, C 0.85%)	7 448	16 711	0.90
Cunife I (Cu 60%, Ni 20%, Fe 20%)	15 597	4 775	0.58
Iron oxide powder (4.96 g/cm³) (Fe_3O_4 92%, Fe_2O_3 8%)	—	31 433	0.75
Platinum alloy (Pt 77%, Co 23%)	30 239	159 155	0.45
Tungsten steel (Fe 94%, W 5%, C 1%)	2 546	5 570	1.03
Vectolite (sintered) (Fe_3O_4 44%, Fe_2O_3 30%, Co_2O_3 26%)	4 775	79 577	0.16

SI UNITS (METRIC SYSTEM)
[See Also Conversion Factors]

SI BASE UNITS

Unit Name	Plural Form	Pronunciation	Symbol	Quantity
ampere	amperes	am′ pār	A	electric current
candela	candelas	kăn de′ lə	cd	luminous intensity
kelvin	kelvins	kĕl′ vin	K	thermodynamic temperature*
kilogram	kilograms	kil′ō grăm	kg	mass
meter**	meter	mē′tər	m	length
mole	moles	mōl	mol	amount of substance
second	seconds	sek ′ənd	s	time

*Degree Celsius (°C) accepted (°C = K − 273.15). Plural form is degrees Celsius.
**Also spelled metre.

SI SUPPLEMENTARY UNITS

Name	Plural	Pronunciation	Symbol	Quantity
radian*	radian	rā′ dē ən	rad	plane angle
steradian	steradian	stərā′ dē ən	sr	solid angle

*Use of degree, minute, and second is acceptable.

SI DERIVED UNITS WITH SPECIAL NAMES

Unit Name	Plural Form	Pronunciation	Symbol	Quantity	Formula
becquerel	becquerels	be 'krel	Bq	radioactivity	s^{-1}
coulomb	coulombs	koo'lom	C	electric charge	A·s
farad	farads	far' ad	F	electric capacitance	C/V
gray	grays	$gr\bar{a}$	Gy	absorbed dose	J/kg
henry	henries	hen' re	H	inductance	Wb/A
hertz	hertz	h^ərts'	Hz	frequency	1/s or s^{-1}
joule	joules	jowl	J	energy	N · m
lumen	lumens	loo' men	lm	luminous flux	cd.sr
lux	lux	luks	lx	illuminance	m^{-2}·cd·sr
newton	newtons	n(y)u' t^ən	N	force or weight	m·kg·s^{-2}
ohm	ohms	$\bar{o}$m	Ω	electric resistance	V/A
pascal	pascals	pas'k^əl	Pa	pressure or stress	N/m^2
siemens	siemens	$s\bar{e}$' m^əns	S	conductance	A/V
tesla	teslas	tes' lă	T	magnetic flux density	Wb/m^2
volt	volts	$v\bar{o}$lt'	V	electric potential	W/A
watt	watts	wät'	W	power	J/s
weber	webers	web'ər	Wb	magnetic flux	V·s

SOME DERIVED UNITS WITHOUT SPECIAL NAMES

Unit Name	Symbol	Quantity
square meter	m^2	area
cubic meter	m^3	volume
meter per second	m/s	velocity (linear)
radian per second	rad/s	angular velocity
meter per second squared	m/s^2	acceleration (linear)
radian per second squared	rad/s^2	angular acceleration
newton meter	N·m	moment of force (torque)
kilogram per cubic meter	kg/m^3	density
joule per kelvin	J/K	entropy
watt per square meter	W/m^2	thermal flux density

NON-SI units used in specialized fields and those of practical importance will remain in use internationally.

NON-SI UNITS MOST COMMONLY USED WITH SI

Name	Plural	Symbol	Value in SI
minute	minutes	min	1 min $= 60$ s
hour	hours	h	1h $= 3\ 600$ s
day	days	d	1d $= 86\ 400$ s
degree	degrees	°	1° $= (\pi/180)$ rad
minute	minutes	'	1' $= (\pi/10\ 800)$ rad
second	seconds	"	1" $= (\pi/648\ 000)$ rad
liter	liters	l	1l $= 10^{-3}$ m^3
metric ton	metric tons	t	1t $= 10^3$ kg
bar	bar or bars	bar	1 bar $= 10^5$Pa

PREFIXES OF SI UNITS

Name	Pronunciation	Symbols	Amount	Multiples and Submultiples	Definition
exa	ex' a	E	1 000 000 000 000 000 000	10^{18}	one million million million times
peta	pet' a	P	1 000 000 000 000 000	10^{15}	one thousand million million times
tera	ter' a	T	1 000 000 000 000	10^{12}	one million million times
giga	ji' ga	G	1 000 000 000	10^{9}	one thousand million times
mega	meg' a	M	1 000 000	10^{6}	one million times
kilo	kil' o	k	1 000	10^{3}	one thousand times
hecto	hek' to	h*	100	10^{2}	one hundred times
deka	dek' a	da*	10	10	ten times
deci	des i	d*	0.1	10^{-1}	one tenth of
centi	sen' ti	c*	0.01	10^{-2}	one hundredth of
milli	mil' i	m	0.001	10^{-3}	one thousandth of
micro	mi' kro	μ	0.000 001	10^{-6}	one millionth of
nano	nan' o	n	0.000 000 001	10^{-9}	one thousandth millionth of
pico	pe' co	p	0.000 000 000 001	10^{-12}	one millionth millionth of
femto	fem' to	f	0.000 000 000 000 001	10^{-15}	one thousandth millionth millionth of
atto	at' to	a	0.000 000 000 000 000 001	10^{-18}	one millionth millionth millionth of

*These prefixes should generally be avoided except for measurement of area and volume, and for nontechnical uses of centimeter.

TRANSMISSION LINE DATA

Type of Line	Characteristic Impedance (Ω)	Velocity Factor*	Attenuation in dB per 100 ft. at 28 MHZ**
Open two-wire line	400-600	0.975	0.10
Air-insulated coaxial line	50-120	0.850	0.55
Solid dielectric coaxial: RG-58/U	53	0.660	1.90
Solid dielectric coaxial: RG-59/U	73	0.660	1.80
Twin lead	300	0.820	0.84

*Ratio of velocity of wave along line to velocity in free space.
**Assuming no standing waves.

Index